THE SECRET LIFE
OF THE FOREST

RICHARD M. KETCHUM
THE SECRET LIFE OF THE FOREST

Conceived and produced in cooperation with the St. Regis Paper Company

AMERICAN HERITAGE PRESS   NEW YORK

ACKNOWLEDGMENTS

In 1965 the St. Regis Paper Company initiated a series of remarkable advertisements designed to explain the secrets of the forest to the general public and broaden understanding of man's reliance upon the forest community. Most of the illustrations created for that purpose were the work of Jack Kunz, an artist whose sensitive, penetrating knowledge of nature illuminates so many of these pages. His drawings, and those by Rudolf Freund and Bernard Pertchik, reveal with striking clarity the marvelous inner life of the woodlands that make up such an important part of the North American continent. To these artists and to their painstaking, often inspired work, this book owes much.

The publishers are extremely grateful to the St. Regis Paper Company for the full measure of cooperation and the generous assistance it has contributed to this project at every stage of its development.

INTRODUCTION

For most of human history, man's involvement with nature was both intimate and complete. The primitive's wisdom was his accumulated knowledge about the environment of which he was a part; his skills were techniques of adapting to the natural surroundings. Only in fairly recent times has man removed himself from the partnership nature requires, ignoring its incalculable blessings and forgetting the terrors it can hold.

What men have lost, in consequence, is what Dr. Rene Dubos calls a "theology of the earth"—the sacred relationships that link humankind to all the physical attributes of the planet. Centuries ago, when the Greeks began to personalize their deities in order to make the gods' actions more comprehensible and to free the natural world from some of the fears it held, they imagined that those wood nymphs they called dryads were responsible for the forests and were capable of punishing outrages against the trees they guarded. Yet for all their respectful attitude toward the deities of forest and soil and water, the Greeks and other ancient peoples unwittingly brought about the ruination of portions of the environment that sustained them; for the goats and sheep of primitive farmers were as efficient as modern man's earth-moving equipment in destroying vegetation. Much of the Mediterranean basin, and the treeless Scottish moors—perhaps even what is now the Sahara Desert—are testimonies to carelessness or ignorance of certain basic natural laws.

Each passing year brings new evidence that all living things on earth are bound together in a complex yet fragile interrelationship. Indeed, as Professor Barry Commoner writes, "The web of relationships that ties animal to plant, prey to predator, parasite to host, and all to the air, water, and soil which they inhabit persists *because* it is complex."

One of the miracles of life on earth is the cycle that begins with the green plants, which use energy from the sun and combine it with water and carbon dioxide to create food, and at the same time liberate oxygen into the atmosphere. Man, with all the other animals, is the beneficiary of this wondrous process: animal life consumes the plants, making use of the energy stored within them, and breathes the oxygen that has been released by green organisms. Now that man is in a position to dominate all life-forms on earth, one of the great questions that confronts him is whether he will destroy, in his restless, unquestioning conquest of nature, the chlorophyll cycle on which he is totally dependent. As Dr. Dubos puts it, we must learn "to reject the attitude which asserts that man is the only value of importance and that the rest of nature can be sacrificed to his welfare and whims."

Theories in the abstract are not always easy to follow or to adopt, and one of the purposes of this book is to advance understanding of the inner workings and significance of one important and familiar form of plant life —the tree. There is today a growing realization of the perils mankind faces if it does not cease or alter significantly the practices that have despoiled so much of the environment. For too long we have been neglectful or intolerant of the role played by forests in the cycle of life; and one of the most heartening developments of recent times is the expanding effort to renew that working partnership with nature, to use its bounties wisely, to restore what has been taken away, to replenish and to protect—as the zealous Greek dryads did—the resources on which we are so dependent.

No one knows exactly how or when life began in the sea, but some two billion years ago plants similar to algae were photosynthesizing, or using the sun's energy to manufacture food. For another billion and a half years life evolved slowly in the "primeval soup" of the oceans.

Trees, like all other forms of life, trace their origins to the sea. To understand how the first tree appeared on earth, we must look back across unfathomable eons, to a time more than three billion years ago when the planet we live on began to cool off and change from a molten sphere to one that had a solid crust.

As it cooled, a thin layer of granite formed over the fiery interior; the hot inner mass slowly cooled and contracted, ridges were thrust upward to form mountains, molten lava surged up through cracks, and boiling water rose to the surface. As anyone knows who has seen one of the many hot springs that even now gush up out of the earth, this process is still going on; geysers and active volcanoes testify to the searing heat that prevails far inside the earth. Scientists believe that the water in our oceans today was first released by volcanic action as a gas, which formed the primeval atmosphere. When this vapor reached extremely high altitudes, it condensed into water and fell earthward. For a long time, however, because the atmospheric temperature was so hot, it resumed its gaseous form before reaching the planet, but eventually the surface of the earth cooled enough so that water began accumulating upon it in liquid form.

And then, for literally millions and millions of years, it must have rained continuously, the water sweeping minerals down from the rocks and filling the depressions in the earth's face. For as much as half of its total existence, the world was barren of life. No living thing stirred anywhere. And then, perhaps two billion years ago, the miracle of life began, and eventually the congenial ocean environment became a "primeval soup," teeming with minute organic materials. For millions of years these simple, organic compounds—a blend of the gases and minerals that had come from the earth's interior and from the atmosphere—combined and interacted. As they did so, certain distinctive features of what we recognize as living organisms began to emerge; cell membranes were formed, and finally, there developed the method of recombination that we call reproduction.

About 420 million years ago, plants first appeared on land, and the first trees established a tenuous foothold on the rocky surface. These trees had no roots: a tangle of specialized branches lay on the surface in the debris of seaweed and moss from the sea. Eventually, land plants helped decompose the bare rock and create soil.

Some of the organisms that were the ancestors of plants found a means of using sunlight as a source of energy—a process known as photosynthesis, about which we shall see more on page 20—which enabled them to survive and prosper in an inorganic environment. Photosynthesis, the activity that ultimately makes all life possible, permits plants to trap energy from the sun and store it in the form of glucose, or sugar, for their growth. To make one molecule of sugar requires six molecules of water and six of carbon dioxide, taken from the air. When these are combined with the energy from the sun's light, glucose is formed—to be stored within the plant—and oxygen and carbon dioxide are released into the air. As minute organisms became more plentiful, increasing quantities of oxygen began to be thrown off and built up in the earth's environment. Since oxygen in its free form is highly poisonous to most organic molecules, only those primitive organisms that were able to tolerate it could survive, and they gradually evolved a method of using oxygen in a new kind of energy cycle.

Unlike geology or zoology, in which there are fossil remains to tell the story of early life on earth, the beginnings of plant life are lost in the mists of time, and the study of the first plants rests upon conjecture and educated guesses. (Fossils do not appear in any quantity until the beginning of the Cambrian period, which is to say, about 600 million years ago. These remains do not mean that life suddenly appeared at that time—they only indicate that many forms of life, such as plants, did not acquire hard parts capable of preservation until then.) The sea originally occupied a much larger area than it now does, but it was quite shallow; and we can surmise that at some distant time the original plant cells evolved near the surface of the ocean, or in its offshore waters, and that these first cells became the basis for higher plants.

A plant has been defined as a living thing that absorbs in microscopic amounts over its surface all that it needs for growth. Through an unimaginably long and complicated process, the first genuine plant forms began to take shape in shallow sea waters and ultimately moved out of marsh and estuary onto the land, adapting themselves into forms that were capable of surviving on the inhospitable coastlines. We do not know how plants made their transition from water to land, but at some point in time seaweeds or some borderline plant forms must have begun to grow onto the muddy shores and into the air and found a way of anchoring themselves there when the waters ebbed. After leaving the friendlier surroundings of the oceans, plants were exposed to a vastly more complicated environment in which more difficult conditions of survival forced them to adopt more complex forms. Struggling into the unfamiliar dry air, standing up to the rigors of sun and wind, required entirely new mechanisms in order to endure. Even today we can see remnants of the changes that took place: simple, primitive forms are still found alongside the most sophisticated varieties of plants, and plants with all the characteristics of sea organisms are discovered on a mountaintop, far removed from the ocean. Lichens, for example, are the hardiest of pioneers; because they are not dependent upon soil, these remnants of a prehistoric time grow farther north and higher above the timber line than any other plant.

On land, of course, conditions differed markedly from those in the sea, and before plants could grow in any quantity, the soil in which they would live had to be created. We have already seen how gigantic upheavals inside the earth brought elements to the surface; minerals in the molten rock squirted upward, usually following cracks made by earthquakes, sometimes erupting in awesome volcanic blasts, and most of the atmosphere and the water for our planet came from these vents in its crust. The long, slow process of soil-building began on the first rocky surface. As the forces of nature broke up rocks, the sun's warmth caused them to expand, and the working of water created larger fissures in the earth's face. The constant cycle of heating, cooling, soaking, and drying opened little cracks and pockets into which bits of loose rubble and minerals settled. Wind and water swept stony fragments from the mountains, water

10

and ice ground them into smaller pieces, and storms picked them up and deposited them miles from their original location. In land areas near the sea, the deposits of one season were exposed to the air as the floodwaters receded; when the next rainy season came along, the surface was again flooded by water, which carved out new channels in vulnerable soft spots, and as the waters once more lost their force, mud and sand were deposited, forming a layer of rich sediment. The soil mantle deepened with the passage of millions of years, and the individual particles became increasingly finer. Water seeping through the layer of soil carried tiny pieces of mineral-laden dirt downward, gradually creating subsoil. Primitive plants found a foothold in crevices and cracks, and began to spread. There was still no animal life on land, nor could it exist until there was sufficient plant life for a food supply. For millions of years the landscape was a monotonous brown, broken only by the spreading green tide of vegetation; there were no flowers to give it color, no sound of animals or birds.

The most important surviving relics of the primitive plants are mosses and ferns, which give us a clue to how ancient plants solved the problem of reproduction. They developed spores, or microscopic cells, which fell into the water, germinated, and formed male and female sex organisms that combined to create a new plant. Spores are relatively inefficient; it takes thousands of them to form the almost invisible dust on the underside of a fern leaf, and when these are released, only a few of them ever lodge in a place that offers the combination of sunlight and moisture they need to develop.

Eventually other forms of plant life appeared on the earth's bleak crust. With the passage of time debris from all these plants fell to the surface, adding to the depth of the soil; other, more complex plants appeared, and microorganic forms of life began. And somewhere, somehow, the ancestor of the first tree grew from the thinly covered rock. It was probably not much more than two feet tall, it had only a puny stem and no true leaves, but it towered above the other plants of its time and stood up to the wind that swept across the barren

Insects and ancestors of the scorpion lived in dense forests of giant ferns and club moss 370 million years ago.

Plant life piled up in the marshy soil and was compacted by its own weight into coal and oil 300 million years ago.

surface of the world. It was the culmination of hundreds of millions of years of evolution, and it was able to stand on its own, above the other plants, because it embodied a woody structure that supported it in its upward struggle toward the sunlight.

The first forests about which anything is known were made up of the simplest kinds of plants, all of them spore-produced—horsetails, club mosses, and ferns. Relics of these primitive plants still survive today; everyone is familiar with the delicate, lacy ferns that are found in damp places in the woods, growing in the niches of rocks, spreading over the remains of fallen trees. But these are only miniature versions of their ancestors. During the Paleozoic era, between 280 and 425 million years ago, at about the time that reptiles were evolving, ferns forty feet tall and club mosses five feet in diameter and 120 feet in height flourished in swampy lowlands between recurrent invasions of the sea. As these huge plants fell over into the water and slowly turned into peat, they retained their wondrously rich stores of solar energy, acquired through photosynthesis, which were compacted first into lignite and then into coal by the growing weight of millions of years of deposits. Forest grew upon forest, and each in its turn was compressed into seams of coal beneath the surface of the ground. When we burn coal, we are using a fuel made from the sun's energy and stored away in trees more than a quarter of a billion years ago.

We know that the passage of time has produced a steady, progressive change in plants, with new, improved species coming along to supplant the old ones. Although some of the earliest primitive plants survive, they have in the main been replaced, or shoved aside, by more adaptable, or more aggressive, species. And so it was that the early ferns, club mosses, and horsetails were outstripped by a group of plants that had devised a more efficient method of reproducing. These were the first plants to produce seeds, and they represented a true revolution in the green world.

A seed, as we shall see later, is actually an embryo plant, fully equipped with leaves and its own food supply, which nourishes it until conditions are favorable for the plant to germinate and grow. Unlike the spore plant, whose fertilized egg could become a fern only if it found a source of food and if conditions were suitable for immediate growth, seed plants had the power to delay germination—that is, to remain dormant—for months or even years. Seeds made possible the spread of plants from the prehistoric bogs and swamps to the dryer land, to plains and mountains that had never supported vegetation of any kind. And the woody vegetation produced by some of these seeds had more chance of survival in a land environment, since what was required was a combination of rigidity, longevity, and the ability to attain sufficient height to outreach the competitive growth. One of the earliest seed plants was the cycad, a palmlike tree with fleshy seeds and large, fernlike leaves, which may still be seen in tropical forests. It was one of the most abundant plants in the landscape two hundred million years ago, but it was gradually replaced by another type of seed plant whose descendants are among the most familiar on earth today—the conifers.

Breaking through the relatively low seed-ferns and cycads at about the same time the first warm-blooded mammals appeared, the conifers became for millions of years the dominant trees on the planet. Their seeds, contained in distinctive cones, enabled them to dominate the environment and overshadow the spore plants, and to spread into habitats where there had been no previous growth. Today's descendants of those ancient conifer forests—pines, spruces, firs, larches, cedars, cypresses, and junipers—include some of our tallest trees and the oldest living plants. They cover great stretches of the North Temperate Zone, including high mountain ranges; they are our most productive timber trees, supplying three-fourths of the lumber used by man and nearly all the pulpwood used for paper products. But they benefit few other kinds of life, providing no food for man and little for birds and animals. Prevalent as they are, they have been fighting, for millions of years now, a losing battle with yet another type of tree

—the one that is the culmination of plant life on earth—the flowering tree.

This plant, whose multitude of forms are so familiar to us today, was a broad-leaved organism that provided flowers, fruit, leaves, and—when men got around to using them—lumber and an extraordinary variety of other wood products. The first flowering trees, whose covered seeds represented a major advance over the naked seeds of the conifer, made their appearance when dinosaurs ranged the land, and they provided new forms of food for herbivores of all types. Eventually these flowering plants grew almost everywhere that life could exist, and in the form of trees, shrubs, grasses, and herbs, they now supply virtually all the food on which man depends. As a result of the rich profusion of life generated by the flowering plants the earth's flora by sixty million years ago was strikingly similar to what we know today, with the flowering forest and grasslands spreading across the land, pushing back the conifers and other forms of plant life in an evolutionary struggle that still goes on.

Today's flora is the culmination of a process that has been taking place for some billions of years—a process that may be easier to comprehend if we imagine what would have occurred had the story occupied just one week in time. Just after midnight on Sunday, when our telescoped sequence begins, the first microscopic organisms emerge in the seas, and from then until 6 A.M. on the following Saturday they combine and expand until the ocean is full of primitive plant life. Not until 10 o'clock on Saturday night, in our capsule version of time, do the first plants take their tentative steps onto the land. Then the process begins to accelerate: by 3 A.M. on Sunday the great seed-ferns and cycads cover the earth in a forest that formed the great coal beds that men mine today. At 10:30 on Sunday night the flowering plants emerge and begin to dominate the landscape. Finally, less than three minutes before midnight on Sunday, the evolutionary force that has had more effect on plant life than any other in the long history of the world—man—appears on the scene.

Seed-producing plants, including the conifers, appeared about 200 million years ago, with the first dinosaurs.

Deciduous trees—flowering plants that shed their leaves—emerged at about the time the dinosaurs vanished.

If we think of a seed as a box or container with a miniature plant inside, it is easy to understand why those first seed-trees outstripped other forms of plant life that were not so well equipped. The spore plant, of which the fern is an example, requires the presence of moisture before it can reproduce. The spore grows into a tiny green plant upon which sex organs appear; but only if water is present—in the form of rain or dew—can the male cell swim to fertilize the female egg. Thus, moisture must be available at precisely the right time for reproduction to occur—a process that leaves a great deal to chance.

Inside the container of the seed, by contrast, the embryo plant already exists, with a root, a trunk, and one or two leaves. Not only is it ready to start growing the moment conditions are favorable but within the seed are cells containing food reserves of protein, starch, and oil to support the baby plant until it roots itself and begins to draw nourishment from its environment. (An extreme example of the amount of food that seeds store for the plant is the palm coconut: its big shell contains enough food to nourish a young palm tree with leaves up to a foot in length.)

We can visualize how a new tree begins life by thinking of the seed that was formed on the branch of a tree during the past summer. After it drops off, it is blown or carried by an animal or bird some distance from the parent tree, where it lies dormant through the winter. With the coming of spring the soil begins to grow warmer, rains soften the ground, adding to the moisture already present from the melting snows, and a change begins to take place within the seed. The cells of the embryo plant start to divide, the shell or container cracks open, and the root emerges into its new environment. The drawings on these pages permit us to follow that process.

First the root finds its way out of the seed, and as if by instinct, seeks the ground. The hard tip at the end of the root shoot wriggles into the earth, aided by a combined push from the expanding, dividing cells in the growing plant. As the third drawing from the right shows, once the root has made its way into the earth the rest of the plant takes the shape of an arch. The trunk of the plant is bent like a horseshoe while the top still remains embedded in the seed container. This has the effect of reinforcing the upward thrust of the young tree because it is actually pushing with two legs instead of one.

When the arch has sprung and the plant stands straight, naked in its new environment, the unneeded shell drops off, and the tip of the plant begins to grow upward toward the sun while the branches, or roots, below the surface, reach out in search of moisture. As the inner cells of the plant are affected by different intensities and directions of sunlight, certain chemical reactions take place, and the plant begins to grow upward in a spiraling or irregular motion. The tree's life has begun.

The seeds of pine trees are concealed inside the cone, with two seeds on each scale. When the cone opens, the seeds fall to the ground.

Drawings show the seed of the piñon pine four times actual size.

Inside the seed is an embryo tree with leaves, stem, and root point.

The growing embryo splits the seed shell, and the root point emerges.

When the root penetrates the soil, the tree can absorb water and food.

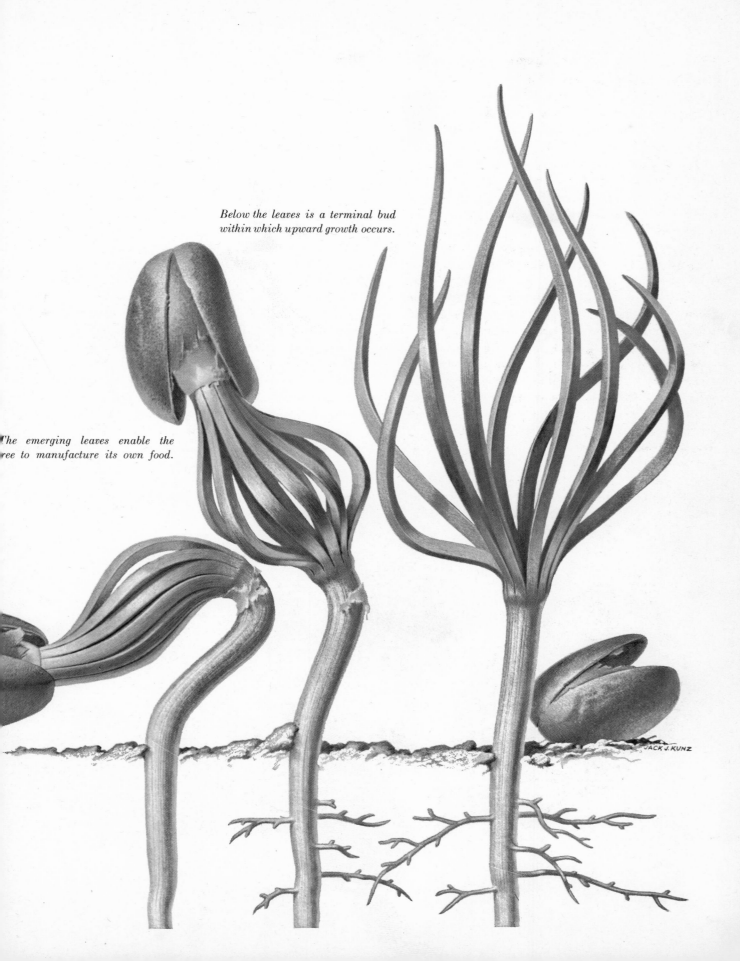

Below the leaves is a terminal bud
within which upward growth occurs.

The emerging leaves enable the
tree to manufacture its own food.

JACK J. KUNZ

Cherries have a juicy pulp around a hard, protective seed-covering. Birds eat the ripening fruit and distribute seeds over wide areas.

Birds and animals eat the oily beechnut; prickly husks stick to the fur of animals, and the seeds may be deposited far from the tree.

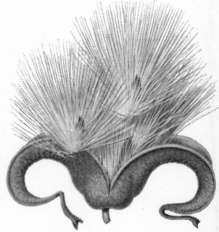

Air-borne willow seeds, embedded in light fluffy tufts, are capable of germinating in a few hours and putting out new green shoots.

Mature cones of the white pine open their scales and expose winged seeds to the winds. These are called gymnosperms, or naked seeds.

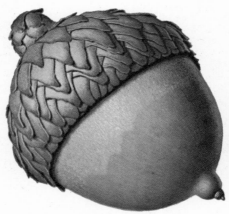

An oak often owes its existence to the squirrel that "planted" it, since acorns must be covered before it is possible for them to germinate.

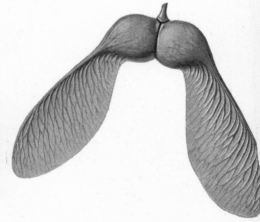

Maple "key fruits" belong to a botanical group known as samaras, or winged seeds. A spinning motion slows the seed's descent.

Perhaps nothing illustrates the trees' ability to adapt to environmental circumstances quite so well as the ingenuity with which they scatter their seeds in order to reproduce. Some seeds simply fall to the ground; some float on the water to find a place to germinate; others are fired like buckshot over surprising distances from the parent tree. Seeds are enclosed in nuts or fleshy fruit, to be carried away by animals and birds; they are concealed within burdocks or sticky seed pods that attach to an animal's fur; they sail through the air on wings, propelled by the winds.

Whatever the means of transport or the size and shape of the seed container, the seed within has the same basic form and purpose: it is the fertilized "egg" of a new tree, seeking a place to take root in fresh ground to perpetuate the species.

Seeds develop from flowers, whose stamens and pistils produce the sperm, or male cells, and the eggs, or female cells. When these are united, a seed is formed. Depending upon the plant from which it comes, the seed may range in size all the way from the tiny, powderlike speck of the rhododendron to the black walnut or even to the huge palm coconut. They have an almost infinite variety of shapes and colors and other characteristics, some of which are indicated by the drawings on pages 16 and 19. Yet there are in fact only two different types of seed: one of them is represented by the pine cone at left, while all the other illustrations show members of the second seed group.

The first, and older, seed-bearing trees are the conifers—typified by such species as the pine and spruce—which make up as much as a third of all the world's forested areas. These trees produce seeds in their cones, which are simply a concentration of scales bearing ovules that become seeds when the cone ripens. Since they have no outer case or protective covering, they are known as *gymnosperms*, from the Greek words meaning "naked seeds." In order for fertilization to occur, pollen is produced in the male flowers, and in the spring of the year these tiny sperm cells are released, to be carried by the wind into contact with egg cells in receptive female flowers or cones. When the cone ripens, its scales loosen or open, and the naked seeds fall to the ground. Although this method of reproduction represented a great evolutionary improvement over the older, more primitive spore plants, it is nonetheless a chancy process, one so profligate that the water of a pond in a pine forest is covered with a film of billions of pollen grains at the time the trees flower.

The other type of seed, which resulted from the trees' evolutionary development of a more orderly and more certain method of reproduction, is the *angiosperm*, from the Greek words meaning "seeds in a container." We have already seen that the first flowering plants began to emerge some sixty million years ago, since which time they have proliferated, adding to the world a rich bounty of flowers and fruits and inexorably pushing back the great coniferous forests. No one knows how the first flowering trees evolved, but botanists imagine that at some remote time insects discovered that the sap exuded by ovules, or eggs, in the female portion of seeds would make a tasty meal when mixed with the male pollen. As the insects returned again and again to this source of food, they stimulated a more systematic form of reproduction than the one that had depended solely upon the vagaries of the winds; but at the same time the insects doubtless bit into many of the eggs, destroying them and preventing reproduction. Gradually, in the course of evolution, plants began to develop a wall, or carpel, around the ovules to keep insects from reaching the eggs, and this protective layer became, in time, a seed covering.

As the first angiosperm plants spread inland onto dry ground, the insects followed their food supply; in pursuit of them came the reptiles that fed on insects. Eventually the reptiles discovered that they could substitute plants for insects in their diet, and they grew larger as they browsed among this more available and reliable source of green food. Plants that had come to rely on insects for pollination devised new methods of attracting them: the reproductive organs were surrounded by bright, showy foliage that could be seen from a distance; flowers developed; and with the passage of time certain plants evolved techniques of attracting particular animals. The purpose was to insure that their regular pollinators—the animals that would best serve the plant, and not chance visitors that raided its pollen and nectar —would be most likely to revisit it. Flowers that are fertilized by bees, for instance, are hardly ever red, since most bees have difficulty seeing this color. The motion of certain flowers in the wind apparently attracts some animal pollinators, while other flowers display patterns of color—dots or lines that converge at the entrance to the food supply to lead the insect to it.

With the development of flowering plants a whole new world of wildlife grew and enlarged—an expanding complex of birds and animals and insects that depended upon the gathering of nectar and that pollinated other plants as a side benefit of their unending quest for food. In the case of certain plants, such as the cherry and apple trees, animals consumed the fruit, and the seed passed unharmed through their alimentary canal to be deposited far from the plant of origin. Small animals that fed on the forest's abundance became the prey of carnivores and thus part of the food chain, which will be discussed later—the vitally important system by which matter from soil and air passes through plants and animals and back to soil and air. It is this system upon which all life depends.

In their urge for survival the seed-bearing trees hit upon countless different devices for carrying pollen from one flower to another, but essentially the methods fall into two main categories. The first is wind-pollination, which requires the presence of light, small, dry pollen grains, easily shaken from the stamens, or male flowers. To receive the tiny bits of pollen that are blown about by the wind, the stigmata of flowers must be long, or feathery, or sticky, or so constructed as to trap the fine dust. All conifers are pollinated this way, as are the poplar, ash, birch, oak, beech, and certain other species. But since this is such a haphazard method, in which a disproportionately high percentage of pollen is wasted, these trees must of necessity produce immense quantities of pollen in order that even a tiny amount will be effective. Scientists have estimated that a single stamen of a beech tree, for example, may yield two thousand grains, while the branch system of a vigorous young birch can produce 100 million grains a year. One pine or spruce cone alone releases between one and two million grains of pollen into the air; in Sweden, which is covered with spruce forests, an estimated seventy-five thousand

tons of pollen are blown from the trees each year.

Most of these pollen grains, which appear to the naked eye like minute particles of dust, fall to earth to become part of the humus accumulating on the floor of the forest, and because the outer wall of the grain is so durable, scientists studying the contents of ancient peat bogs or the layers of mud beneath an old lake often find pollen grains intact many milleniums after they fell from the trees. The steady rain of pollen sank to the bottom of the lake each year, one layer piling up on top of another to form strata. In a bog of this type the botanist can make a boring with a special cylinder and bring up from the bottom a core that contains samples from each layer. By examining these he can discover the nature of the vegetation, the changes in forest types, and the variations in climatic conditions that took place long before the advent of recorded history.

A far more sophisticated method of pollination than that of broadcasting it on the winds is the one in which the plant compensates for its own lack of mobility by having animals work for it. As a general rule, the plants that rely on this technique have found a way to supply the animal with food and warmth and protection while it collects pollen and nectar, and the pollen is located where some portion of the animal's body will rub against it. (We know that certain animals are capable of communicating to others of their kind the existence of a food source: bees, for example, that have been out searching for nectar return to the hive and fly or "dance" in such a way as to indicate the direction, distance, and richness of food.) And at times the lure of shelter is as important to an insect as the attraction of food.

Typically, animal-pollinated flowers have attractive petals, scent, and nectar; they are generally conspicuous and have many-seeded fruits. By contrast, the wind-pollinated plants usually have flowers that are small and inconspicuous; they have numerous hanging male catkins from which the pollen is easily shaken or blown. The wind-pollinated trees—of which the willow and alder are typical—flower in the early spring before the new leaves come out, and before insects are on the wing, so that they can shed their pollen before the emerging foliage can prevent its flight.

Fruits of the tropical mangrove tree germinate on the branches and form pointed, stiltlike roots. These float to shore and lodge in the mud.

Seeds of the witch hazel tree are fired into the air when the pods dry, contract, and squeeze out the seeds with a loud snapping noise.

The seagoing seed of the coconut has a nearly waterproof covering. The coconut "milk" nourishes the plant until it can establish roots.

The thin-shelled pecan is another nut that can float long distances and still take root. Wild trees produce nuts that are rich in food value.

If it is not buried by a squirrel, the black walnut may float on the water to a new location. Walnuts contain more protein than milk.

Trees like the apple depend on deer and other animals to eat and spread their seeds, which pass unharmed through the digestive tract.

In the spring, when melting snows and surface water pour off the hillsides into dark, swollen streams, the earth begins to stir with an unseen, inner force, like some primeval monster coming to life. As the sun's warmth penetrates the ground, the woods take on a misty, light-green veil, delicate new leaves begin to unfold, and the trees commence their annual cycle of growth. But the forest is more than earth and water and new green plant life: inside each tree, invisible to the eye, a highly efficient organism is functioning with all the precision of a superb machine.

Despite all this outward evidence of new life and growth, the curious fact is that the vast bulk of a mature, healthy tree is actually dead. The only visible parts that are truly alive are the leaves, buds, flowers, and seeds. Concealed within the tree, its living cells, which may comprise only a tiny percentage of the total, are constantly at work—in the roots, absorbing moisture from the ground; in the cambium sheath under the outer bark, transporting the food needed by the tree and enlarging the trunk; in the tips of twigs, elongating them; and in the leaves, working at the process of photosynthesis by which the tree combines moisture with energy from the sun's light and transforms this energy into nutrients and carbon dioxide. The "dead" cells are made into structural material, the wood of the tree.

A tree is an intricate mechanism whose interrelated parts must constantly work in harmony for the organism to survive and prosper. In ancient times men used to speak of the four elements—earth, fire, air, and water—of which all substances were thought to be composed. If we think of the living tree in relation to these four elements, we see that earth is what supports it and supplies it with nutrients. Fire, which represents the light and heat of the sun, provides the energy that it converts into food. Air supplies the gases it consumes in "breathing," the moisture it absorbs through its leaves. And water, which it must have in great quantities, is the basis of protoplasm, the living contents of the cells.

We know a great deal more than the Greeks did about the forces of life. But with all our knowledge, no one fully comprehends the miraculous process, known as photosynthesis, by which trees and other green plants use energy from the sun to transform elements into food for themselves at the same time that they release oxygen into the atmosphere. This is what makes life on earth possible, and it is the most important function performed by green plants of all kinds.

One of the first men to learn something of this activity was a seventeenth-century Flemish physician, J. B. van Helmont, who conducted an experiment proving that a tree increases in size by means of substances taken from the air. He began by planting a willow tree in a tub of soil that he had carefully weighed, and for five years he watered the tree and watched it grow. At the end of that time he took the willow from the barrel, weighed it, and discovered that it had gained more than 164 pounds, while the weight of the soil in the tub remained virtually unchanged. All that Van Helmont had added was water, which meant that the tree had manufactured its own growth from some source outside the soil.

Photosynthesis, which means "putting together with light," is a manufacturing process that takes place inside the cells of plants, in tiny bodies called chloroplasts. A single leaf of a tree contains millions of these chloroplasts, and each chloroplast has many layers of chlorophyll, which is the green coloring matter of plants. When photosynthesis takes place, water that has been absorbed by the tree's roots is carried to the leaves, where it comes in contact with the layers of chlorophyll. At the same time, air, which contains tiny amounts of carbon dioxide, has also penetrated the leaves. When sunlight hits the leaf, it supplies the energy for a chemical reaction in which the water is broken down into its separate elements of hydrogen and oxygen. They combine with carbon dioxide in the chlorophyll to form glucose, a simple sugar, which the tree will use as a source of food. Meanwhile, the excess oxygen resulting from the process is released into the atmosphere by the leaves. This is the way the process is expressed as a chemical equation:

$$6CO_2 + 6H_2O + \text{energy} \rightarrow C_6H_{12}O_6 + 6O_2$$
carbon   water   expended   glucose   oxygen
dioxide

This indicates that six molecules of carbon dioxide combine with six of water, in the presence of the sun's energy, to form one molecule of glucose, which is a carbohydrate, and six of oxygen. As we have seen, the oxygen goes into the atmosphere while the glucose is carried to the other parts of the tree as nourishment. From it, some 95 per cent of the tree is eventually formed, as a result of other chemical actions. The creation of organic matter out of thin air and water is a magical trick that makes possible all life on this planet.

How the tree utilizes the product of photosynthesis is a function of its internal structure. The entire substance of a tree consists of cells, those tiny units of life. Everything about the mature plant—its size, shape, and texture—is attributable to the character of the cells, which shape the leaves, flower petals, fruit, and bark and determine the nature of its wood. The cells' remarkable versatility makes it possible for trees to endure vastly different climates, temperatures that range from far below zero to the searing heat of the desert, water supplies that vary from fresh to brackish, violent changes—from torrential downpours to droughts that shrivel the vegetation. Each cell that makes up the tissue of the root, stem, and leaf of a tree is like a little bubble, endowed with life and with all the potentialities of the entire plant. The cell is surrounded by a thin elastic wall and is filled with protoplasm, which carries its life properties. It also contains all the information or instructions that will govern its own reproduction and determine what type of cell it will become: root, bark, or leaf.

A characteristic that distinguished the first trees from more primitive plants was their ability to develop stems and roots and to stand erect, and the process through which this occurs is the continual division and multiplication of cells. The protoplasm of the cell absorbs water and pushes against the elastic cell-wall, making it

The hard tip of a root consists of living, growing cells (shown in a dark color at the bottom of this cutaway view) that push through the soil. Behind the protective cap are minute root hairs, which are single-cell projections that absorb water and dissolved minerals from the earth and start them on their way to the leaves. A large, mature tree may have hundreds of miles of roots anchoring it to the soil, but most of them consist of dead, woody matter.

as taut as a full balloon. Every cell along the chain of the stem does the same thing and internal pressure builds up, pushing outward and upward, as if thousands of individual balloons were being blown up, and eventually increasing the diameter and height of the tree. In effect, the trunk and leaves of the tree hold themselves erect by means of the bulging cells inside their skin. We have all seen what happens, on a smaller scale, when a garden plant does *not* have enough water: without the bulging cells to keep it erect, the whole plant wilts and collapses like a deflated balloon.

How complex is the interrelationship of all the tree's parts is evident when we examine the process of growth, by which tissue is formed from the raw materials of water, minerals, and carbon dioxide. For a time, the tree seedling we observed on pages 14 and 15 is almost impossible to distinguish from other plant shoots in the forest. But after about three or four weeks in the case of hardwood trees and by the end of the second year in the case of conifers, the seedling begins to exhibit a true stem and other treelike characteristics. The growth that has taken place has already begun to form wood.

Having sprung from the sea, trees have never lost their dependence on water, and while the leaves would appear to be the parts of the tree that gather most of the moisture, the real water-collecting mechanism is in the tiny, nearly invisible hairs behind the root tips. The tip itself is hard, and as it probes and burrows into the soil like a curious finger, the delicate hairs wrap themselves around individual grains of soil and absorb moisture and dissolved minerals from them. The root tips, at which Charles Darwin marveled, saying that they behaved almost "like the brain of one of the lower animals," push out in all directions beneath the surface,

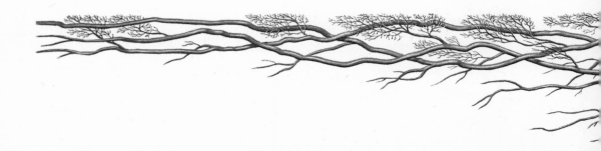

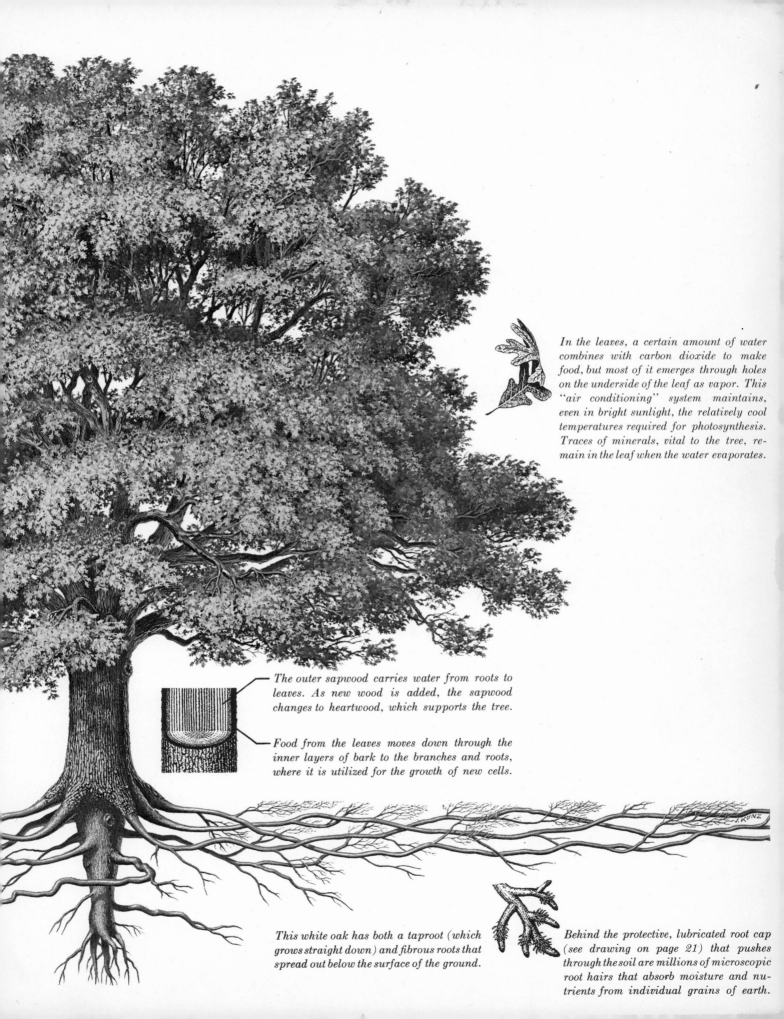

In the leaves, a certain amount of water combines with carbon dioxide to make food, but most of it emerges through holes on the underside of the leaf as vapor. This "air conditioning" system maintains, even in bright sunlight, the relatively cool temperatures required for photosynthesis. Traces of minerals, vital to the tree, remain in the leaf when the water evaporates.

The outer sapwood carries water from roots to leaves. As new wood is added, the sapwood changes to heartwood, which supports the tree.

Food from the leaves moves down through the inner layers of bark to the branches and roots, where it is utilized for the growth of new cells.

This white oak has both a taproot (which grows straight down) and fibrous roots that spread out below the surface of the ground.

Behind the protective, lubricated root cap (see drawing on page 21) that pushes through the soil are millions of microscopic root hairs that absorb moisture and nutrients from individual grains of earth.

moving through the earth in a corkscrew motion, turning aside when they strike a rock. Behind them the main roots continue to lengthen and expand, developing as much surface as possible for the absorption of water. Since the minute root hairs live for only a few days, it is essential that the tree grow new roots to replace them. Apart from its need for a root system large enough to anchor it in the ground and support it, the tree must develop roots that will reach down into the subsoil, where the mineral matter is often richer than it is near the surface. (As a general rule, trees that grow in swamps and in wet bottomlands are shallow-rooted; on high, dry ground, where the roots have to reach deep into the subsoil for water, trees are usually deep-rooted. The latter are well anchored and less likely to be blown down by windstorms than shallow-rooted trees are.) As the roots branch and rebranch, just like the limbs of the tree above ground, they increase in size, and because they are tapered, function somewhat like a wedge being driven through the ground. By the time a tree is full-grown, the underground root system is enormous; a mature oak tree, for example, has literally hundreds of miles of roots to tap the soil's resources in an endless quest for water. Each drop is collected by the root hairs and passed along, from one cell to the next, up the trunk and to the leaves, and in such a way that none of the precious moisture and minerals collected by the roots leaks back into the soil.

The immediate beneficiary of all this root development is the tree, of course; but the soil itself profits in many ways, too. Gradually, the tiny root hairs reach out to so many particles of earth that the soil becomes firmly tied into place, with the result that it is capable of resisting the erosion that occurs from the action of wind and rain. As the roots work their way into the soil, they take the sun's energy deeper and deeper beneath the surface and at the same time break up the earth for other living things—animals, insects, worms—that will enrich the soil. Eventually, the action of bacteria decomposes dead plant and animal matter, creating a rich humus and releasing carbon dioxide into the air. We will see more about this process on pages 49 to 51.

At the same time the roots are spreading out, the branches above ground are growing, and they and the leaves need a continuous stream of moisture from below. Unlike the closed circulatory system of an animal, in which a fixed amount of blood is recirculated from one part to another, the tree's internal plumbing might be compared to a kerosene lamp. The wick of the lamp is anchored in liquid and sends moisture up to the top, where it is consumed. Similarly, a tree is an ever-flowing wet system that must be maintained at all times; once the flow of moisture ceases, the tree dies, just as a lamp goes out when the kerosene is gone and the wick is dry.

In order to grow, the underground roots must be fed from the leaves of the tree; roots, of course, receive no sunlight and are not able to photosynthesize. This has

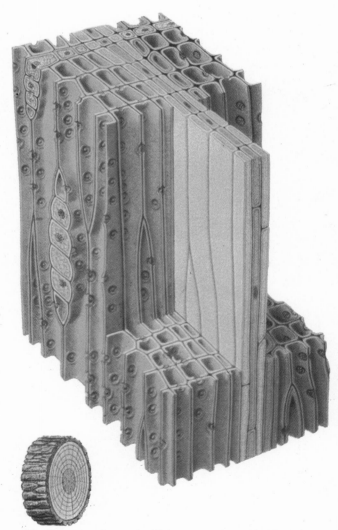

The thin green layer of cells known as the cambium divides to form bark on the outer side and wood on the inside.

to be done for them by the leaves; but the feeding of the roots must also be achieved in such a way that the upward flow of water from root to leaf is not disturbed. Inevitably, as the tree becomes taller, its upper and lower extremities—the tips of branches and roots—grow farther and farther apart. In order to adjust to the facts of growth, and at the same time maintain the two-way flow of water essential to its needs, the tree has to grow internally, to expand. This is accomplished by the thin sheath of cells known as the cambium layer.

The trunk of a tree is formed by a special growing layer situated between the wood and the bark—the cambium. A soft, sticky film, the cambium consists of a microscopic layer of cells that continue to divide throughout the life of the tree. It performs the unusual double function of adding both to the wood that builds up in the inner body of the tree and to the bark that develops on the outside. In the spring of the year the cambium begins to work, splitting off wood cells on the inside and bark cells outside. In some trees, bark has a tendency to peel or slough off, but the inner wood builds up continuously. That portion of it that is formed in the spring, when the tree has plenty of water and its growth is most vigorous, consists of large, light-colored cells; in the summer and early fall, when the tree grows more slowly because there is less moisture available to it, the cells are smaller and thicker-walled, so they appear to be darker. This is what creates the rings that may be seen in a cross section of the trunk when a tree has been cut down: each pairing of light and dark creates an annual ring, which enables us to estimate the age of the tree—one for each year of its life.

The important business of conducting water up and down the length of the tree is handled by an intricate system of tubes, which develop from the expanding cells. Those that bring water and minerals up from the roots are inside the cambium, in the sapwood portion of the tree, and they are known as the xylem. The ones that compose the food track—that is, the tubes that carry food from the leaves down to the roots—are called the phloem, and they are located outside the cambium, on the bark side of the tree. In this highly efficient plumbing system, water and dissolved minerals enter through the root hairs underground, are transported upward through the xylem, and infuse the leaves, where they are converted by photosynthesis into sugar and protein. This food then travels from the leaves downward through the phloem to the roots, to support their continuing growth. In addition to serving as conduits for moisture and food supplies, the xylem and phloem store food for the tree's use. When the phloem is cut off from the water supply inside the tree, it begins to die and forms the protective layer of bark on the outside of the tree. The xylem also ages, and as it does, becomes impregnated with a substance known as lignin, which binds together the woody cells in a process called lignification. This is what produces the tough, fibrous substance we know as wood.

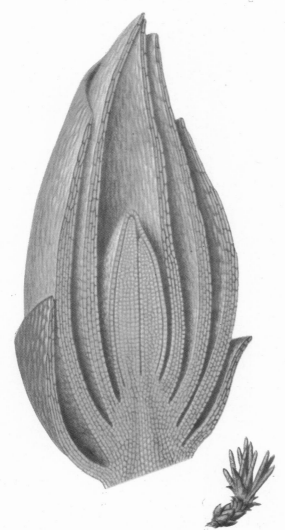

When cells at the base of a leaf bud (above) divide and elongate, they form a new twig behind the developing leaf.

Concealed within the skin, or outer bark, of a mature tree, as this illustration indicates, are four distinct layers, each of which plays its particular role in the life of the plant. The outer bark acts as a suit of armor against the outer world, warding off insects and disease, protecting the inner tissues against damage from storms or extremes of temperature. Just inside it is the inner bark, or phloem, which plays out its limited life span as purveyor of food from the leaves to the roots before dying and adding its substance to the outer bark.

The third layer inside the skin is the remarkable cambium sheath—a thin film of tissue that throws off cells to the phloem and xylem and is the agent by which the trunk expands its diameter. Although appearances may suggest otherwise, a phenomenon of tree development is that the trunk does not increase in stature; all upward growth takes place in the tips of branches and roots—the trunk increases only in diameter and girth as the cambium adds new layers of cells. If a man drives a nail into a tree trunk at eye level, years later the nail will remain in the same position even though the height of the tree may have increased considerably.

Beneath the cambium is the xylem, or sapwood section—the cells through which minerals and water move up to the branches and leaves and in which the tree's reserve food supply is stored. Finally, at the center of a large, mature tree is the core, or heartwood—its main support. This is formed when the older water-carrying tubes of the sapwood are no longer needed. As they become inactive and fill with wastes or resin, they harden to create the heartwood, just when the tree needs added strength for support. The tubes and fibers within the heartwood are inert and no longer essential to the tree's survival except as reinforcement; this is the reason a hollow tree can go on living for years, or as long as the living tissues outside the core continue to fulfill the tree's vital functions.

The outer bark insulates the tree from extreme heat and cold, helps to keep out rain, and protects the plant against insect enemies.

The phloem conducts food from leaves to the rest of the tree. Eventually, it becomes part of the tree's bark.

The cambium layer produces new bark and new wood annually in response to hormones, called auxins, that stimulate the growth of cells.

Sapwood is the pipeline for water moving from roots to the leaves. When its inner cells lose their vitality, they turn into heartwood.

Heartwood is the central, supporting column of the mature tree. Although it is dead, it will not decay or lose strength as long as a tree's outer layers remain intact.

Our most accurate method of reconstructing the past is by examining the historical record. Archaeologists consult the artifacts of earlier civilizations, historians the archives of another era, paleontologists the fossils of prehistoric animals. The forester has the unique record of the rings.

We have already seen that each year of a tree's growth is recorded by the addition of another ring to the cross section of its trunk; it is as if a completely new, tight-fitting cylinder had been superimposed around the circumference to chalk up another year of life. If the ring is wide, it indicates that growing conditions that year were favorable, with ample sunlight and moisture. If it is narrow, it suggests an abnormally dry season, or crowding by the tree's competitors in a dense forest, or some other unfavorable circumstance that reduced normal growth. By comparing tree rings, scientists have been able to compile weather charts reaching back hundreds of years, and the rings of trees cut centuries ago can throw an illuminating light on the activities of man as well.

Until they were rediscovered in 1888, the Mesa Verde cliff dwellings of the Pueblo Indians had been silent

1904
The loblolly pine tree is born.

1909
*Undisturbed, the tree grows rapidly.
Abundant moisture and light
produce broad, evenly spaced rings.*

Two-thirds typical size

28

1914
Wider "reaction wood" rings on one side show that something pushed against the tree, making it lean.

1924
Now growing straight, the tree is crowded by others whose crowns and roots compete for water and light.

1927
The surrounding trees are harvested, making available ample nourishment and light for the tree's growth.

1930
A forest fire scars the tree, but in successive years new wood builds up and eventually covers the wound.

1942
Narrow rings like these were probably caused by a dry spell of more than one or two years' duration.

1957
Another series of narrow rings may have resulted from insect attack— perhaps by larvae of the sawfly.

and deserted for hundreds of years. About A.D. 1100 the Pueblos had moved into the most advanced stage of their civilization, which lasted for nearly two hundred years, but sometime in the fourteenth century their strange cliff cities had been evacuated, never to be inhabited again. By examining the tree poles used in constructing their buildings, scientists concluded that a terrible drought had forced the inhabitants to depart about the year 1290: the rings of the trees, preserved by the dry climate, revealed that they had all been cut before that date.

Tree rings can be "read" in other ways, too, for legible within the silent wood is documentary evidence of the major events of the plant's life, as the illustration on page 29 shows. When it was born, how it grew, how it withstood the competition from neighboring trees and survived natural catastrophes—all these, as well as its life span, can be discovered in the rings.

The shape of a living tree may also suggest how it has fared in its struggle for existence against weather and other species. In the deep woods, leafy crowns form the canopy, or "overstory"; beneath it, the lower branches that are cut off from the source of light often die and drop off in a process of self-pruning. Where openings are made in the forest, little trees suddenly shoot up, and the hardiest ones reach the sunlight first. Those that lag behind are called suppressed trees, and unless nature or the lumberman gives them a chance to grow, they develop slowly or frequently die.

Usually it may be assumed that a healthy tree's wide branches and dense foliage have shaded out its rivals. Its symmetry may have permitted the sunlight to reach more of its leaves, producing more food and building a stronger, taller specimen. Or in certain instances the species itself, like those shown on this page, has developed a particular shape that enables it to exist under extremes of cold, wind, and moisture. The ubiquitous black spruce, which is found from Labrador to the west coast of Alaska, may grow to be a hundred-foot tree with a straight, slender trunk in the relatively congenial southern section of Canada but at the northern limits of its range, where it is whipped by arctic wind and ice, the mature tree may be no more than a twisted shrub ten or twelve feet tall.

From left to right, the distinctive silhouettes are those of the baldcypress, the sugar maple, the paper birch, a stunted black spruce (at bottom), and the balsam fir.

The bark of a shagbark hickory breaks up into distinctive plates, loose at one or both ends, as the tree matures.

White birch is often called paper birch because of the texture of its bark, which peels off in paper-thin strips.

Until the ponderosa pine is eighty or a hundred years old its bark is black; the mature tree takes on the color shown.

The mangrove's leathery bark helps it adapt to its unique habitat—near or in salt water in tropical regions.

Sycamore bark has a tendency to peel off in large, thin, brittle plates, revealing lighter-colored areas beneath.

The tough, fibrous bark of the redwood may be as much as a foot thick. It is very resistant to fire and insects.

One of the distinguishing features of trees, by which we identify one variety from another (especially in wintertime, when the deciduous trees have lost their leaves), is bark. Since new bark forms inside the tree through the action of the cambium layer, what we see on the surface is the aging bark. The expanding tree splits and cracks the bark as if it were trying to break out of its skin; the old bark weathers, sloughing off the outer layers, changing in appearance as the tree matures.

Bark cells are quite short-lived, and the outer bark of a tree is, in fact, dead—like the inner heartwood. As the walls of the dead cells lose their vitality, they become impregnated with a corklike substance and eventually fill with air, which enables them to function as a waterproof coating, an insulator, and a fire wall for the thin layer of living cells inside. The dimensions of bark vary from tree to tree: some never develop a thick protective coating, whereas the bark on some of the giant sequoia trees in the Sierra Nevada Mountains is two feet thick. This has saved many of them from damage by forest fires, for the material is virtually fireproof. Not only its thickness but the rich tannin content makes it too formidable a barrier for insects to penetrate (so far as is known, none of these old trees has died of disease or insect attack). The barks of certain trees contain substances that are distasteful or actually poisonous to animals, and this is one way living trees fight off beetles and other insects. Often, when a tree is bruised or its bark smashed, insects suddenly appear, attracted by the smell, to infest the wood, and they sometimes kill the tree.

Early settlers in the thick eastern woodlands used to "girdle" trees to kill them; by chopping around the circumference of the tree, through the bark and the cambium into the sapwood, they cut off completely the downward flow of food to the roots and interfered with the upward motion of water and minerals, slowly strangling the plant. The same thing happens when a porcupine gnaws a circle around the trunk of a tree; these destructive animals frequently destroy whole groves of maple, beech, and other valuable specimens.

What makes it possible to use bark in identifying trees is the characteristic manner in which the trees respond to the stretching process. As the illustration of the sycamore trunk indicates, its bark is tight and stretches very little; patches of the old bark flake off, and the new, yellow material underneath gives the tree a peculiar mottled effect. By contrast, the fibrous skin of the redwood is quite pliable and maintains a generally uniform appearance throughout the height of the tree. Strips of the paper birch—a bark used by the Indians for the hulls of their canoes—peel off to reveal a fresh, new white layer underneath.

That part of the tree so familiar that we use it as a convenient form of identification is in many respects the most mysterious. The consuming activity of every natural community is the business of obtaining food, without which the living organism perishes. So, in the sense that the leaf is the tree's source of food, all its other parts—roots, trunk, and branches—are subordinate, serving only as structural support or as the purveyors of raw materials. Not only trees but man and all the animals are dependent upon green leaves for food. In addition to plants in the form of vegetables, man consumes various kinds of meat or fish or fowl that have fed on green plants of one sort or another; the hawk that eats a mouse is devouring a small herbivore that fed upon plants. All flesh, as the farmer knows, is grass.

Before animal life could begin, there had to be enormous surpluses of plant food to sustain it, and for hundreds of millions of years the leaves of green plants have been producing that food. On what a scale may be imagined if we take the example of one large sugar maple tree—the producer of the leaf illustrated on page 35. If all the several hundred thousand leaves of a mature sugar maple could be spread out on a flat surface, they would probably cover half an acre, and to sustain the tree, they present approximately that much surface to the sun. During the course of a summer day a square yard of leaf surface manufactures somewhere in the neighborhood of one gram of sugar per hour, or a total of a pound and a half per month during June, July, and August, when the sun is at its brightest. That means that our mature sugar maple is capable of producing more than 3,600 pounds of food—nearly two tons—in one summer season through the action of its leaves.

As the drawing indicates, the leaf has two more or less flat surfaces that are interlaced with an intricate system of veins. These veins are the continuation of the xylem and phloem tubes in the rest of the tree, and they serve as the conduits through which water and nutrients from the soil are carried to the leaf and through which food manufactured in the leaf is taken to the rest of the plant. The leaf's surface is covered with a waxy, transparent coating that serves the double purpose of holding the internal structure of the leaf together and preventing its innards from being damaged or destroyed by wind and rain. This outer skin is a highly effective water jacket; inside, the leaf's photosynthetic tissues are surrounded by a water supply that may comprise as much as 90 per cent of the content of a young leaf, and it is the added task of the coating to prevent this moisture from evaporating.

Beneath the outer layer, suspended like clusters of green balloons, are the cells that make up the main photosynthetic tissues of the leaf. Placed on end, in order to receive and trap the maximum amount of sunlight throughout their length, are the so-called palisade cells, which occupy about half the leaf's depth. These are filled with chloroplasts, the tiny bodies that contain chlorophyll, the link between the sun and life on earth, which gives the leaf its green color. Below the palisade cells are loosely packed, spongy cells.

The air containing carbon dioxide used in photosynthesis enters the leaf through minuscule openings, or air valves in the surface, called stomata. Each of these is a kind of mouth, between two lips (the word "stomata" comes from the Greek word for mouth), which may be seen in the drawing on page 37 as rows of yellowish, lip-shaped openings in parallel rows. Our maple leaf probably has as many as ten thousand stomata on its outer surface. When photosynthesis occurs, the air comes in through the stomata and moves through the spongy layer into the smaller cells of the palisade tissue. During daylight carbon dioxide is removed from the air inside the leaf by photosynthesis, and the oxygen released by this action diffuses out of the leaf through the stomata while the sugar that has been manufactured accumulates in the plant, making possible the growth of protoplasm.

Photosynthesis depends upon several factors. The most favorable conditions for its occurrence are when the temperature is about 70° F. and the light moderate and somewhat diffused. A good supply of water is essential, and the soil's fertility determines whether the tree's body will receive an adequate source of minerals for healthy growth. At night, when there is no sunlight to provide energy, photosynthesis ceases, but the tree continues to respire (a tree has no lungs, of course, so it does not actually breathe). This is another important process in the tree's metabolism since it involves the oxidation, or burning, of sugar manufactured by photosynthesis, with a release of energy. We may compare

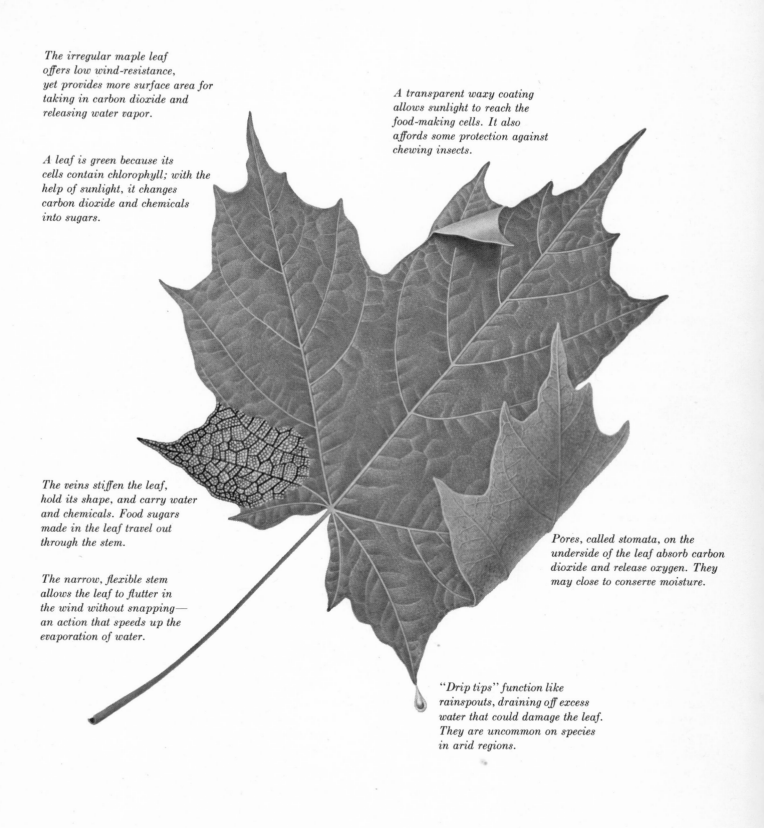

The irregular maple leaf offers low wind-resistance, yet provides more surface area for taking in carbon dioxide and releasing water vapor.

A leaf is green because its cells contain chlorophyll; with the help of sunlight, it changes carbon dioxide and chemicals into sugars.

A transparent waxy coating allows sunlight to reach the food-making cells. It also affords some protection against chewing insects.

The veins stiffen the leaf, hold its shape, and carry water and chemicals. Food sugars made in the leaf travel out through the stem.

The narrow, flexible stem allows the leaf to flutter in the wind without snapping— an action that speeds up the evaporation of water.

Pores, called stomata, on the underside of the leaf absorb carbon dioxide and release oxygen. They may close to conserve moisture.

"Drip tips" function like rainspouts, draining off excess water that could damage the leaf. They are uncommon on species in arid regions.

the chemical equation for respiration with the one for photosynthesis on page 21 and see that it is the direct opposite. Respiration is expressed this way:

$$\underset{\text{glucose}}{C_6H_{12}O_6} + \underset{\text{oxygen}}{6O_2} \rightarrow \underset{\substack{\text{carbon}\\\text{dioxide}}}{6CO_2} + \underset{\text{water}}{6H_2O} + \underset{\text{released}}{\text{energy}}$$

Since respiration, which consumes food or glucose, takes place at all times, night and day, it might seem that the tree would respire itself to death, because photosynthesis, which manufactures the food, occurs only during daylight. But fortunately, photosynthesis produces glucose at ten times the rate that respiration consumes it, and what is left over is used by the tree for growth. Curiously, photosynthesis is less efficient when the weather becomes too hot, and in time of drought it may stop altogether. When this occurs, the tree's life is endangered, because the warmer it gets the more the tree respires, and the more it consumes the reserves of food that are no longer being made by photosynthesis.

Another essential life-function involving the leaf is called transpiration. We have observed how vital water is to the tree: it is an essential ingredient of all the tree's tissues, including the dead ones; the activity of photosynthesis takes place in water; the plant's food is built from it; and minerals are carried to the plant in water. In the process of transpiration the water absorbed by the roots is pushed up into the sapwood of the tree and on to the leaves, some of them hundreds of feet above the ground. As in photosynthesis, the energy required by transpiration is furnished by the sun, and the operation is another of those miraculous achievements of a tree. To take an extreme example, consider the problem involved in lifting water to the leaves of one of California's giant redwood trees, some three hundred or more feet above the ground from roots that may be another one hundred feet or more beneath the surface. To raise water to a height of four hundred feet without the aid of some gigantic pump would appear to be impossible, yet the trees do it—and at a rate of flow estimated at 150 feet an hour in certain species. What pulls the sap up is the result of the action taking place in the leaves. When water is lost by evaporation or as a consequence of photosynthesis, a shortage is created to which the water in the roots responds. Since the molecules of water have cohesiveness, a tendency to stick together, there are at

all times, in the thin vertical veins of the tree, continuous strands or columns of water in the sapwood—just like the column of mercury in a thermometer. When a shortage occurs in the leaves, a tremendous pull causes the water to move up through the veins of the tree's conduction system from root to leaf.

Although the tree is capable of shutting the leaves' stomata and preventing loss of water, it can only do so by keeping out carbon dioxide at the same time, thus preventing photosynthesis. In order to set up a balance between these two requirements, the tree's stomata open early in the morning, making photosynthesis possible when the water supply is most likely to be plentiful. Then, during the heat of the day, around noon, the stomata begin to close, and by sunset they are shut for the night.

Since the sun is the source of their energy, trees have developed various means of exposing their leaves to it. The limbs of a tree are generally spaced at equal intervals around the trunk, and each type of tree has a certain angle at which it puts forth its limbs. The same rules are followed by the branches, by the twigs that emerge from the branches, and by the leaves that emerge from the twigs: each of them develops at the same angle the limbs make with the trunk. On some trees—among them the maple, ash, and horse chestnut—the branches, twigs, and leaves are arranged opposite each other. Other trees have a spiral or alternate arrangement: no two branches or leaves emerge opposite each other.

Different species behave differently in the wind, to take advantage of every opportunity to face their leaves toward the sun. One of the most distinctive actions is that of the quaking aspen, whose thousands of tiny leaves seem to shimmer and flutter simultaneously from top to bottom of the tree in the slightest wind. Oak leaves tend to wave up and down on their heavy branches; the long, flexible branches of the willow sway and bend in the breeze. As John Muir wrote so poetically: "The winds go to every tree, fingering every leaf and branch and furrowed bole; not one is forgotten; the Mountain Pine towering with outstretched arms on the rugged buttresses of the icy peaks, the lowliest and most retiring tenant of the dells; they seek and find them all, caressing them tenderly, bending them in lusty exercise, stimulating their growth, plucking off a leaf or limb as required, or removing an entire tree or grove, now whispering and cooing through the branches like a sleepy child, now roaring like the ocean; the winds blessing the forest, the forests the winds, with ineffable beauty and harmony as the sure result."

A leaf consists of a number of separate parts, which may be seen in the illustration of the maple leaf on page 35. The entire flattened portion of the leaf is called the blade. The stalk, or stem, which produces and supports the leaf, is known as the petiole, the nature of which varies from tree to tree: some are long, some short; most of them are round, although in poplars and aspens they are usually flattened. The extension of the petiole that runs through the center of the blade to its tip is called the midrib, and the veins emanating from either side are, as we have noted, extensions of the tree's water system.

On pages 38 and 39 we will see more about the variety of leaf types and shapes that exist; what is worth noting here is that all of them fall into two major categories— simple or compound. When the leaf blade is one piece, as it is on an oak or maple, it is called a simple leaf; if the blade is divided into small leaflets, as on the locust, ash, or hickory, it is termed compound. The very number of these small blades makes for many more opportunities to catch the sun, and compound leaves are consequently more efficient than the simple variety. Another advantage the compound-leaved tree has over the simple-leaved one is the relative ease with which it can repair damaged blades; it can survive many setbacks because its small leaves and other parts are so easily replaceable. Trees with thick primary stems, big twigs, and large leaves are more easily injured by frost; damage to a bud or leaf means that another large bud or leaf must be constructed, and in this type of tree the process is considerably slower than it is in one with compound leaves. The coconut palm, for instance, has enormous leaves that may take as long as five years to develop fully, and replacing one of them is a major task.

Many of our trees are broad-leaved, and we seldom think of the gymnosperms or conifers as having leaves, but that, of course, is just what their needles are. As we can see from the illustrations on the next two pages, an incredible variety of leaf forms exist in the North American forests.

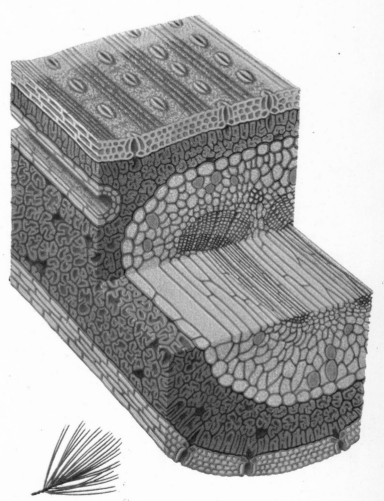

The needles of a conifer tree are actually leaves, which manufacture sugar by combining water from the roots with carbon dioxide from the air. This process, called photosynthesis, is made possible through the action of chlorophyll in the leaves and the utilization of energy from sunlight. The sugar compounds are passed along to living cells; these combine sugar with oxygen, transforming the sun's energy for the life processes that enable them to grow.

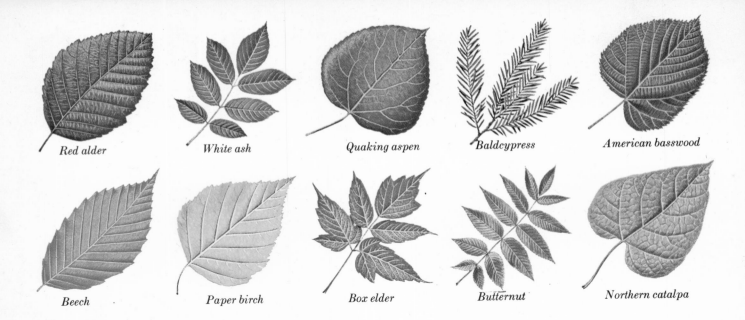

Red alder   *White ash*   *Quaking aspen*   *Baldcypress*   *American basswood*

Beech   *Paper birch*   *Box elder*   *Butternut*   *Northern catalpa*

Although nearly 1,200 forest trees are found on the North American continent, almost all the genera are represented by these fifty-five leaves. The most unusual example here is the ginkgo, a fan-shaped leaf found in no other flowering plant, from a unique tree that is the only survivor of a family that flourished in the age of dinosaurs. The sassafras leaf illustrated is one of three different-shaped leaves that often grow on a single branch of that tree; and the white-oak leaf is but one of the seventy-five or eighty separate species of oaks found in North America. In both angiosperms and gymnosperms a continuity of form extends through the leaves within a tree genus, with the result that oaks and maples, as well as pines,

of different species normally resemble one another.

Usually the needles are arranged in clusters or bundles on pine trees, while other evergreens have different arrangements of their leaves. On spruces, for example, the needles appear singly, along the branches and twigs; on the hemlock and balsam fir they are narrow, flat, and with parallel edges; in some cedars they are small and scaly, with overlapping edges.

Whatever shape they take, however, the purpose of the leaf is to manufacture food for the tree, and the lobes, leaflets, and jagged edges are there to help evaporate the water used in photosynthesis, to reduce wind resistance, or to function as "drip tips" for shedding excess water.

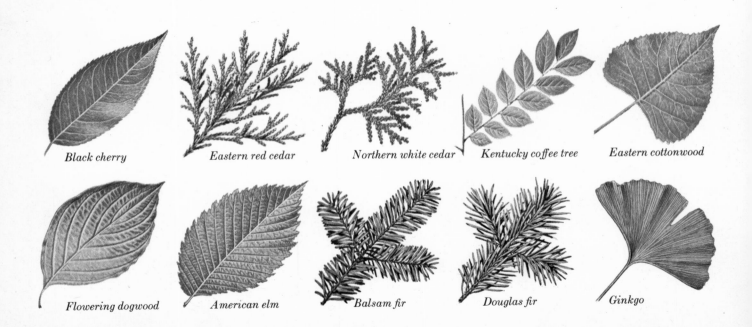

Black cherry   *Eastern red cedar*   *Northern white cedar*   *Kentucky coffee tree*   *Eastern cottonwood*

Flowering dogwood   *American elm*   *Balsam fir*   *Douglas fir*   *Ginkgo*

Hackberry   Hawthorn   Eastern hemlock   Shagbark hickory   American holly

American hornbeam   Horse chestnut   Rocky Mountain juniper   California laurel   Black locust

Honey locust   Southern magnolia   Sugar maple   Red mulberry   White oak

Osage orange   Cabbage palmetto   Pecan   Persimmon   Eastern white pine

Balsam poplar   Eastern redbud   Redwood   Sassafras   Giant sequoia

Red spruce   Sweet bay   Sweet gum   American sycamore   Tamarack

Tulip tree   Black tupelo   Black walnut   Black willow   Pacific yew

Unlike an animal, which can move about freely in search of food and water, the tree, once its seed has germinated, is locked in place and is at the mercy of certain external forces. In addition to sunlight, the most important of these environmental factors—over which the plant has no control and to which it must adapt—are the water supply, temperature range, soil characteristics, and certain surface features. Of these, the most critical is water, for water and life are inseparable.

All the tissues of a tree—those that are dead as well as the living ones—contain water. Before the seed can sprout, there must be water; a young leaf consists almost entirely of fluid; the moisture content of the trunk may be as high as 50 per cent of its total volume. Apart from the water embodied in the cells of a tree, all its vital processes, as we have seen, occur in the presence of this life-giving fluid. Some idea of the prodigious amounts of moisture required by a tree is apparent from the fact that 55 grams of water are required to manufacture 100 grams of cellulose; yet at the same time the tree is adding those 100 grams to its weight, it is losing—through transpiration—the incredible amount of 100,000 grams of water.

Because of this staggering consumption of moisture a single acre of forest land, during the course of an average growing season, may produce between three and four tons of usable wood (not including the new growth that makes up its roots, branches, leaves, and flowers). To manufacture those three or four tons requires some four thousand tons of water, which must be absorbed by the roots, trunk, branches, and leaves, providing all those members with nutrients from below ground before the moisture vanishes into the air as a consequence of transpiration. Such immense infusions of water enable the tree to grow, but the same water is then released into the air after serving the tree's purposes. Nor is the consumption of water on such a scale confined merely to trees: it takes nine hundred pounds of water to produce a pound of dried alfalfa; five thousand pounds to produce a pound of wheat. The corn plant shown on the opposite page uses approximately fifty gallons of water during its one-hundred-day growth period. To demonstrate the importance of moisture, an experiment was made in Utah on two plots of land—one of them irrigated, the other without irrigation. The dry land produced twenty-six bushels of corn per acre, while the

irrigated plot yielded fifty-three bushels to the acre—more than twice as much.

All the available water in the environs of our planet is contained in a closed, constantly recirculating system. In other words, the amount of moisture accessible to all forms of plant and animal life now and in the future is fixed—which explains why it is so essential to preserve its purity. It exists in three states: as a liquid, on or beneath the surface; as a vapor, in the air; or as a solid, in the form of snow or ice. The main reservoirs are the oceans, which cover nearly three-fourths of the planet, and the sun provides the energy that keeps this system in constant movement. Water from the earth's surface evaporates and rises into the warm air above; then this warm, moisture-carrying vapor travels upward into the atmosphere where it is cooled. In the process it is transformed into mist, taking the form of clouds, and as it is cooled further it condenses into larger drops of water, which eventually fall back to the earth as rain or snow. It is estimated that some 100,000 cubic miles of moisture is drawn up into the atmosphere in a year's time; most of it returns to the oceans, but about 35 per cent reaches the land in one form or another.

When clouds move across the land and deposit moisture, several things may happen. Some of the water hits the ground and runs almost immediately into streams and rivers that carry it rapidly toward the sea. Water that runs off in this fashion is of almost no value to the land community; to be of maximum use, moisture must be delayed on its path to the sea, and this is one of the principal benefits of plant life. Vegetation slows down the runoff, absorbing much of the moisture and making possible a more leisurely progression back to ocean or sky. Some water becomes part of the structure of plants, some moves through the plant slowly and evaporates into the air, some sinks into the soil to join the huge underground reservoirs beneath the earth's surface. Once in the ground, it is absorbed by the probing roots of plants, to be returned again to the air through transpiration. (The root systems of plants far smaller than trees can be extraordinary networks for the gathering of water. In *The Web of Life* John H. Storer cites a study in which a single plant of winter rye grass was grown in a box that contained less than two cubic feet of dirt. During the course of four months the plant grew twenty inches above the ground and developed fifty-one square feet of visible surface. But beneath the surface 378 miles of roots and 6,000 miles of root hairs had formed to support the plant.)

Some of the snow that falls on a forested mountainside sticks on the crowns, or uppermost branches, where it may lodge for days. Exposed there to sun and wind, part of the moisture is lost by evaporation (as much as 50 per cent may disappear during periods of fair weather). But the rest of the melting snowflakes fall to the ground and pile up in a blanket under the trees. Shaded from the sun, the snow may remain there for weeks or even months, slowly melting and seeping through the forest's accumulated litter into the soil. As it does so, the humus soaks it up like a sponge, and the water collects in crannies of earth and rock, picking up tiny particles of minerals and gradually working its way down to the water table. Finally, the mountain soil brims with water, some of which begins moving downhill like a sluggish underground river, pushed on its way by new supplies of melting snow. A good soil is capable of retaining huge amounts of underground water for the nourishment of its vegetation; poor soil, by contrast, is much less likely to hold moisture.

En route to valleys and plains below, little rivulets of water spurt out of the hillside as springs, creating new streams that follow the shortest, easiest path downhill to join thousands of others in their return to the sea.

When rain falls, it enters the ground more quickly, passing through the humus and penetrating deep into the soil until it reaches the level of the underground water table. (During periods of drought, when the soil is especially dry, most of this water may be absorbed immediately by the parched soil and never reach the water table.) Once the underground reservoir is filled to capacity, additional water that reaches it moves through the water table to the nearest stream. When the weather is warm and ground water is drawn upward by the trees, to be evaporated by the sun, the soil beneath the forest cover begins to dry out, until the moisture is replaced by another rainfall, or until the snows of winter cover the ground again.

Trees have a twofold problem with respect to water: they must obtain the requisite amount of moisture for their growth, but they also have to conserve enough of it for later use, against a time when supplies are limited. Although they gather most of their water from the soil, tapping the reservoirs of underground moisture, plants also obtain water from the air, where it exists in vapor-

A corn plant uses prodigious amounts of water— as much as fifty gallons during its hundred-day growth period. Its roots probe deep into the topsoil to collect available moisture from grains of earth.

ous form. In areas of high humidity trees lose less water by transpiration than they do in drier regions, and they accumulate moisture from the surrounding moist air. A good example of the result is the spectacular growth of stands of redwood and Douglas fir in the coastal mountains of California and Oregon, which is made possible by the omnipresent mist and fog. In an area bordering the dry coast of southern California, where no rain falls from May to October, tomatoes, beans, and other vegetables grow all summer without irrigation. In the spring they obtain moisture from the soil, but when the ground water is exhausted, they rely entirely upon dew and coastal fog.

Different types of trees have adapted in a variety of ways to the supply of moisture obtainable. Some, like the baldcypress, larch, and red maple, are able to tolerate excessive amounts of water (the cypress is capable of growing even when its roots are submerged). At the other extreme are trees that can exist where the annual rainfall is less than ten inches—the level below which desert conditions prevail. Some of them, like the evergreens, are able to survive because they lose a minimum of water through transpiration; others, like the cacti, store moisture in their tissues against long, dry spells; and still others have highly developed root systems for locating water wherever it is to be found. But these are the exceptions: the great majority of our trees live in habitats where the annual rainfall is between twenty and forty inches.

Long before man made his appearance on earth the location and distribution of forests had been determined by climatic factors. As we might expect, tree growth is limited or prevented by drought conditions and by extended periods of freezing temperatures, which means that forests are usually found where the annual rainfall exceeds fifteen or twenty inches and where the period free from frost is between three and four months. Where it is too dry, grasses prevail, or there is desert; where it is too cold, we find tundra or ice fields. The earth's great forest zones are in areas where temperature and moisture conspire to produce favorable growing conditions.

Throughout most of the Northern Hemisphere, the western land areas are favored by prevailing winds; they gather the rains that come off the oceans, blown by storms circling the globe from west to east. Perhaps the greatest geographic misfortune of the United States is the fact that the Cascade and Sierra Nevada mountains run from north to south so close to its western coast. These towering ranges catch most of the life-giving water that sweeps in off the Pacific, with the result that the lands on their eastern slopes are semiarid, suitable mostly for grazing. Almost the whole western half of the United States lies in the "rain shadow" of mountains, where the land can be farmed only if it is irrigated by water trapped on the mountain peaks.

Although it appears on no political map of the United States, one of the most important geographic boundaries we have is the twenty-inch rainfall line that runs north and south, virtually through the middle of the country. It has been called the "disaster line" because of the devastating droughts that have occurred to the west of it. East of the line is a countryside of humid agriculture and deciduous and coniferous forests; west of it there is a wide swath that extends from the Canadian border all the way to Mexico—several hundred thousand square miles—where there are almost no trees.

As with trees, so with man: a source of water has always been one of his primary concerns. That vast area between the twenty-inch rainfall line and the western slopes of the coastal ranges was known to generations of pioneers as the "Great American Desert." For the same reasons that few trees grew there, no major settlements were established until most of the country was well populated. Throughout the world prehistoric communities grew up around a good source of water, near a river, lake, or spring, and it is worth reminding ourselves that the deserts of the Near East are spotted with ruins of what were once great civilizations. Originally situated where they had ample supplies of water, they vanished when man and his animals destroyed the vegetation in the hills where their water originated.

A second environmental factor that exerts a profound effect on trees is temperature. Since they depend for so much of their growth on adequate supplies of moisture, trees develop at a greatly reduced rate, or not at all, in the winter months, when the ground is frozen and little water is obtainable. In continuously frigid regions, like the Arctic, the permafrost—about which conservationists have expressed such concern since the great Alaskan oil strike in 1968—keeps the subsoil continually frozen, preventing trees from developing an adequate root structure. In that fragile environment no forests exist, only shallow-rooted plants and lichens. A large proportion of the continent of North America is subject to cold winters, when the temperature is too low for photosynthesis to occur, and the two types of trees respond to this situation in quite different ways. A great many of our native flowering trees are broadleaf, or deciduous, which means that they shed their leaves at the end of the growing season. By doing so in the fall of the year, before the supply of ground water is frozen, they reduce transpiration and the resulting water loss that would otherwise take place through their leaves. With no foliage on the trees, no photosynthesis takes place, and with respiration reduced to a minimum, water increases inside the tree during its dormant period. By the time the buds are ready to open in the spring, before the leaves re-emerge, the water supply within the tree has built up to a maximum, and the soil is full of water from the winter melt. Then, when the leaves reappear, transpiration and respiration commence, and the tree's reserves of water gradually decrease as the ground loses its moisture with the onset of hot, dry weather.

By contrast, the other type of tree—the conifer, or evergreen—achieves reduction of transpiration because its tiny leaves, or needles, have a minimal surface through which respiration can take place. The needles also have a nearly impervious waxy coating that limits moisture loss further, with the result that evergreens accomplish both transpiration and photosynthesis at a slower rate during the summer months. But by retaining their "leaves" through the winter, they compensate for the relatively slow summer activity by carrying on a certain amount of transpiration and photosynthesis during warm winter days.

Curiously, high temperatures are not in themselves injurious to plants; only if the water supply gets too low, or if the rainfall is less than ten inches a year, are trees adversely affected. This is why we find grasslands or desert—but no trees—in regions where the rainfall is too meager to support larger, thirstier plants. Low temperatures, on the other hand, can have effects other than closing off the water supply: unseasonable frosts are extremely hazardous to the fragile flowers and fruits of trees. As anyone who reads of the loss of a California or Florida citrus crop realizes, late spring frosts are the constant enemy of the fruit grower.

In addition to water supply and temperature, other external factors affecting trees are the character of the surface and the soil in which they live. Certain trees thrive in acid soil, which results from heavy accumulations of decayed organic matter; others prefer alkaline soil, created by an excess of minerals. Whatever its tolerance for soil conditions, any tree's growth rate will be retarded if it does not obtain mineral nutrition in sufficient quantities. Some species prosper in direct sunlight, others tolerate shade; what is important is that the proper amount of light be available. In general, trees are extremely competitive for light and have devised various techniques for outstripping their rivals. Anyone entering a thick stand of mature hemlock or spruce is immediately conscious of how dim the sunlight is on the forest floor; the canopy of lush growth can shut out the sun's rays to a point where the ground is virtually barren of other forms of vegetation.

In addition to sunlight, wind plays an important role in the development of trees, affecting the direction and amount of their growth and causing them to increase their rate of transpiration or water loss. When a tree bends with the wind, the newer cells of sapwood on the upper side are stretched while those on the lower side are compressed. If the prevailing wind is strong and constant, the tree frequently develops a permanent crook in the opposite direction. On the wind-swept peaks of high mountains it is impossible for any plants except lichens to exist, and in areas where wind constantly causes the soil to shift, only those plants survive that have developed the technique of seeding quickly and anchoring themselves securely before they are blown away.

Additional factors determining the character of growth are the degree of slope on which a tree is rooted and its exposure to the sun. Typically, northeast-facing slopes are cool and moist, and this is where one finds beeches, hemlocks, mosses, and ferns, along with the particular birds and animals that depend on this kind of vegetation for food and shelter. Southwest-facing slopes are usually dry and tend to support such trees as hickories and oaks, as well as various grasses. In the Rocky Mountains, for example, the Douglas fir populates the north-facing slopes, and the valleys below are covered with blue spruce; on the southern side are great stands of ponderosa pine. And along the Sierra Nevada range, where the western face is wet with moisture sweeping in off the Pacific, the redwood and sugar pine thrive; on the dry, eastern slopes of these mountains are piñon pine and juniper.

Soil is the very stuff of existence, and its creation and well-being are matters of life and death to all plants and animals, who depend upon it for survival. It has been estimated that nature may take five hundred years to create an inch of rich topsoil—an indication of how tragic and irretrievable is the waste when hillsides and meadows are not protected against erosion.

As we observed in the first pages of this book, the surface of the earth once consisted entirely of rock and water. By examining the unhurried but irresistible process by which rock is transformed into soil, it is understandable why it took so many millions of years before plant life was able to get a start on earth. The drawings on these pages reveal in quick, visual form the infinitely slow, nearly invisible method of converting stone into soil.

The first step in the weathering of rock is the action of water: countless millions of raindrops fall on the rocky surface, softening it, wearing it away, slowly dissolving the minerals it contains. Extreme heat and cold contribute to the activity by causing the rock to crack and fragment. In the high mountains and northern lands exposed to glaciers, pieces of rock break off and are pushed along, to be ground into smaller chunks by moving snow and ice. A titanic example of what now happens on a smaller, more confined scale occurred during the great ice ages, when snow piled up to form continental ice masses hundreds of feet thick; these immense sheets pushed down from the north as far as the Ohio River, gouging the bedrock, scraping and pulverizing everything in their path. A glacier one thousand feet thick exerts a pressure of thirty tons per square foot at its base, and in New England the ice sheets tore at the hard crystalline rocks that lay close to the surface, shoving the topsoil away and depositing it off the coast of Connecticut to form Long Island. When the ice retreated for the last time from New England, the rocky surface was strewn only with boulders; not for hundreds of years did a thin layer of topsoil cover the terrain, and even today, a Vermont farmer may turn over as much stone as soil when he harrows a field. The ice advanced across Canada, leaving much of it barren, rocky ground, then scooped out the basins of the Great Lakes from old river valleys, carrying that fine soil mixture southward to deposit it in layers that were as much as three hundred feet deep in some places. In the Middle West the glaciers encountered no mountains or hard rock; there, like a giant bulldozer, the ice leveled the tops of hills and filled the valleys, crushing and grinding rocks to fine powder and mixing the old surface soil with this and with minerals brought up from deep in the subsoil.

As the icecap moved slowly down across the European land mass, trees of the Temperate Zone—which once grew as far north as the Pole—were shoved southward ahead of the advancing ice until they reached the impassable barriers of the Alps and the Pyrenees. Then, since the climate had become too cold for seeds to lodge and germinate, many species of trees disappeared altogether in Europe, leaving that continent with relatively few varieties. But in North America, where the great mountain ranges ran north and south and did not block the glacial tides, most species survived the ice ages, enabling us to enjoy the wealth of trees we have today. More than nine hundred species of trees have been found in North America above the Mexican border, and in just one area—the Great Smoky Mountains National Park—there are nearly as many types of trees as there are in Europe.

The cumulative effect of snow, ice, wind, and rain is to wear away the rocks into increasingly smaller particles. Rock is the principal ingredient of soil, and its gritty mineral grains are essential to sustain plant growth. Yet on some barren stones plant life of a sort gets a start. Those durable organisms—the lichens—can survive on the surface of a rock even though almost no moisture is available, and once established, their fibers begin to secrete an acid that dissolves minerals in the rock. They literally eat their way into it, opening up tiny fissures into which moisture forces its way, freezes, and cracks off additional bits of rock. As one generation after another adds substance to the stone, behind the lichens come more delicate plants, mosses, ferns, and grasses, taking advantage of tiny footholds in which to set down roots. Eventually tree seedlings catch hold in this thin carpet of plant life, the soil slowly builds up on top of the rock, and a complex of roots works away beneath the surface, spreading out in a widening mat, prying into every crevice, holding the soil ever more firmly in place, and preparing a rich seedbed capable of holding moisture and of providing a foothold for larger forms of plant life. But before there can be rich topsoil —what E. J. H. Corner calls "the gold in the basement of the forest"—another essential process must take place. This occurs in an unseen world where the lower forms of animals are at work with fungi, bacteria, and roots.

The action of water dropping on rock softens and erodes it, gradually dissolving the nutritious minerals inside.

Lichens establish a foothold even though almost no moisture is present and open up minuscule cracks in the rocks.

Mosses, grass, and other small plants invade the crevices of the rock, and a layer of vegetation slowly builds up.

Eventually the rock is covered with enough soil and decayed matter to allow tiny tree seedlings to take hold.

A mature natural forest is not unlike a human community, in which individuals of all ages, sizes, shapes, and differing characteristics go about the business of living and growing old and dying in a myriad of ways. In these wooded lands a rich diversity of plants is struggling to root and grow toward the light, while great numbers are dead or dying, breaking up in various forms of decay. And the forest floor is the best indication of this perpetual waxing and waning of life; it is a mass of rotten branches and tree trunks, strewn with fallen leaves, through which new seedlings and saplings are beginning to emerge.

In such a forest there is an excess of growth; over the years twigs, leaves, flower parts, pollen, fruit, seed pods, buds, and bits of bark have fallen to the earth, covering the forest floor with a thickening litter that gradually becomes part of the humus beneath the trees. Giant hulks of trees killed or split in two lean against others or lie on the ground. Trees that have somehow lagged behind in the struggle for light maintain a tenuous foothold, riddled with insects and rot, ready to topple over at the slightest push. Ancient decaying stumps are mute evidence of forest giants whose trunks and branches have long since disintegrated; huge hollows, in which the remnants of roots persist, mark the site of a fallen tree whose root ball was torn from the soil in the violence of its crash to earth.

There are many reasons for a tree's death, almost all traceable to some external force—fire, drought, insect damage, extreme cold, disease, or simply lack of food or light. Lightning, attracted to the tallest specimens, fells some; ice storms coat the trunk and branches and bend trees to the snapping point; high winds tear off branches and blow down trees; hurricanes and gales take down whole swaths of growth, leaving a desolate, twisted mass of wreckage in their wake. Some trees become top-heavy and tumble over from their own weight, or are blown down, knocking apart neighboring trees and smashing branches all around, leaving wounds and broken surfaces that attract insects and invite disease. Healthy specimens are able to resist most invasions of insects or disease, but insect borers, woodpeckers, and tiny creatures work away at every damaged tree, making "sawdust" of the supporting heartwood, and fungi hollow it out so that it becomes an easy victim of windstorms. Fire, of course, is the forest's worst enemy. Even if a tree is not consumed by a forest fire, the intense heat may kill it or make it an easy prey for insects.

As it does to all living organisms, old age comes to trees. Like animals, certain kinds of trees have fairly predictable life spans: a gray birch is considered old when it is forty, whereas a sugar maple may live to be five hundred, and some giant sequoias are estimated to be more than three thousand years old. Unlike animals, however, trees do not age uniformly; as we have noted, the inner cells of a tree trunk may be dead for years while the growing extremities continue to push out into the environment, until they reach the outer limit to which water can be brought to the leaves and food to the roots. When they are old, trees have difficulty respiring, their new growth is not as vigorous as it once was, and the activity of the cambium cells is much reduced. This breakdown in the tree's vitality has visible effects: the leaves become smaller, more and more dead branches are evident, and damage to the bark or limbs is not so easily repaired, since an old tree lacks the recuperative powers of a more vigorous plant. The annual rings of an old tree are narrower, partly because the cambium layer is less active, but also because the tree finds it ever more difficult to provide moisture for its various parts.

But the death of a tree in the forest community is not the end; it is part of a marvelous cycle that prepares riches for the future.

When a big tree topples and is uprooted, huge chunks of soil may be torn up, naturally "plowing" the earth and opening it for the action of millions of small organisms.

Green leaves provide food for the growth and sustenance of plants.

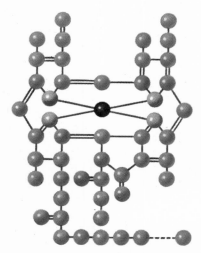

This is part of a chlorophyll molecule, needed for photosynthesis.

Squirrels live and feed in trees and bury their seeds in the ground.

The fox devours small mammals that eat others or feed on plant life.

Insects eat plants and aid in the process of decay, renewing the soil.

Toads live on the forest floor and feed on the small creatures there.

The root network supports and nourishes trees and prevents erosion.

Burrowing through the soil, worms open channels for air and water.

Insect larvae attack dead wood, hastening the process of decay.

Birds contribute to forest life by consuming huge numbers of insects.

Fungi do not photosynthesize, but consume dead plant matter for food.

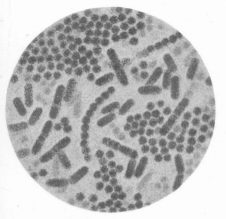

Decay bacteria work on dead matter and release mineral content to soil.

One of the most remarkable characteristics of the flowering forest is its capacity to maintain itself. It is distinguished by the fact that it provides a rich food source for various forms of animal and plant life and by the fact that its remains do not accumulate permanently. There is an inevitable piling up of materials on the forest floor; in an acre of woodland as much as two tons of waste, including dead insects and animal droppings, as well as the debris of branches and leaves from trees and other plants, falls to the ground every year. But these great quantities of refuse are slowly broken down to form organic material through the work of hordes of soil creatures that attack it. This is the environment of the fungi, which can exist in darkness and which, with bacteria and an almost infinite variety of invertebrate animals, live in and exploit the remains of vegetation, eventually reducing it to fine humus. All the minerals that have been absorbed through the roots of the living tree and used in the building of plant tissues are discarded into the wastebasket of the forest floor and transformed so that they can be absorbed by the roots of other plants.

The forest community is a fantastic complex of organisms that live together in the same environment, all of them dependent upon and interacting with the others in some fashion. When the debris from trees falls to the ground, it is attacked first by shallow-boring grubs, which begin to work into it, eating as they go. Then termites and deeper borers appear on the scene, cutting up the dead material into minuscule pieces. They are assisted in the task by small armies of beetles, ants, grubs, centipedes, snails, slugs, and other creatures. Water seeps into the crevices, and fungi of all sorts attach themselves to the tree's remains and hasten decomposition. The wood becomes softer as a result and is soon a place where bacteria can thrive.

Earthworms work away at tiny pieces, taking them

into their network of tunnels below ground, where they soften them by bacterial action and excrete the remains. Their importance as an agent of decomposition can be seen in the fact that the earthworm population of one acre of forest floor is capable of eating eighteen tons of debris in a year's time, mixing it with the soil and leaving tunnels into which water, air, seeds, and bacteria can penetrate. There may be millions of worms to an acre of forest land, and they crush and dissolve leaves and woody fragments in their digestive tracts, mixing those remains with tiny mineral particles they have picked up from the soil itself.

Fungi grow on the dead material, reducing even the bark of trees to a consistency that animals can eat. The excretion and slime from this process become the stuff on which bacteria grow and multiply. Tiny parasitic worms—the nematodes—propagate in the forest trash, consume minute bits of it, and add their excretions to the humus. Predator insects—spiders, scorpions, centipedes, and others—devour the minute soil creatures and add their waste products to the growing humus. The untiring little shrew, only two or three inches long, which must eat at least once an hour to maintain its incredible activity, runs about searching for insects and worms. The mole, in quest of the same prey, burrows through the upper layers of soil, working its path through as much as three hundred feet of tunnels a day.

But the chief agents in transforming litter into the dark, spongy material known as humus are the fungi. These nongreen plants, or plants without chloroplasts, do not photosynthesize like the green plants. They are capable of living in darkness, and they are scavengers that exist on the remains of green plants. They are able to decompose lignin, which is bacteria-resistant, and they reduce the fallen giants of the forest to smaller pieces, making them more susceptible to attack by insects. The fungi—which most of us recognize in their fruiting form as mushrooms or toadstools—decompose the substance on which they are growing by secreting enzymes through their walls. At the same time, they absorb water through their walls for their own growth. The mycelium, or vegetative part, of a fungus grows in rich, deep humus, and it requires a great deal of rain or ample water in the soil. The fungi are most likely to be seen, in fact, after a heavy rain that follows a prolonged dry spell; typically, one finds thriving colonies of toadstools in dark, moist corners of the woods, particularly after the autumn rains have ended a dry summer.

In the deciduous forests of the East, where there is adequate moisture and a generally moderate climate, the leaves that fall in the autumn are decomposed rather rapidly by the small organisms that inhabit the woods—sometimes within a month of the time the leaves have come off the trees. Indeed, autumn is a period of intense activity on the part of all small animals, which are laying in supplies of food for the long winter of hibernation.

The surface of the forest floor may appear to be a lifeless carpet of refuse and dead leaves, but immediately below the top layer are the decaying remains of other seasons, occupied by a busy community of tiny organisms. Deeper still, the humus is honeycombed with the passageways and burrows of insects, worms, moles, and the roots of plants, which make it a protective, insulating cover for the soil and a heavy sponge to check and absorb moisture, preventing rain from running off too quickly.

Growth and decay is the eternal cycle of life in the forest; neither is possible without the other. All the organic matter that has been locked up within a mature tree for dozens or even hundreds of years is released and returned to the soil intact when the tree dies and decomposes. These chemicals are neither reduced in quantity nor is their quality impaired, so the process of growth and decay is one of complete rejuvenation. How that rejuvenation is accomplished is largely the work of the lowly fungi and the unseen bacteria, which break down foliage and wood from trees, and muscle and bone and flesh from animals. The decay bacteria perform one of the most essential functions in our environment by making possible what is known as the carbon cycle. Carbon dioxide exists in the atmosphere, from which, as noted earlier, it is drawn into the leaves of plants as part of the process of photosynthesis. The oxygen molecules are released, and the carbon is converted into the body tissues of the trunk, roots, leaves, fruit, and seeds. When the tree dies, it is decomposed by decay bacteria and fungi and consumed by the millions of forest animals. In the latter process the carbon is absorbed directly into the animals' tissues. In other words, carbon that was originally removed from the atmosphere by photosynthesis is not returned to the atmosphere; instead, it is deposited in the bodies of a multitude of tiny animals. When these creatures die, decay bacteria will feed upon their remains. If they are eaten by predators, which may in turn be devoured by larger animals or birds, the residue of carbon from the tree is passed along intact to each successive predator, adding its bulk to the carbon already in the predator's system. Not until the last link in this food chain dies do the bacteria go to work on the remains. Then they bring about a remarkable chemical transformation by which the complex protoplasmic substances of the animal's body is once again reduced, or broken down, to inorganic, simple forms. By working on the remains of plants or animals, the decay bacteria return the carbon to the air, completing the cycle and making it possible for other plants to draw upon it again for a new round of growth.

Our atmosphere, it is believed, contains only enough carbon dioxide at any one time to support the earth's plant population for forty years. So the part played by bacteria in returning carbon to the air is crucial; without it, all the carbon dioxide would be withdrawn from the air, and vegetation would disappear from the planet.

The dead mouse and the disintegrating leaf will be removed from the forest floor by action of the decay bacteria, which destroy muscle, bone, and other animal tissues as well as foliage and bark of dead trees. They reduce protoplasmic substances to simpler components like carbon dioxide and return them to the atmosphere.

The fleshy tentacles at the end of its nose help the star-nosed mole feel its way underground. Nearly sightless, the mole spends most of its time in a world of darkness, tunneling through humus, preying on earthworms and insects, and opening channels in the soil for penetration by water, air, other creatures, and bacteria.

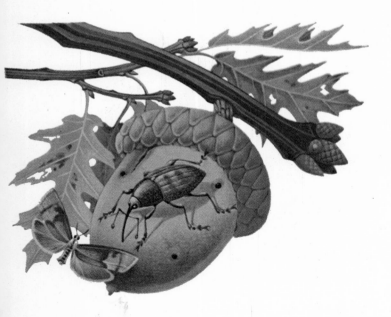

When we think of the word "fruit" the image that comes to mind is of one of the familiar fleshy fruits produced by trees—apple, orange, pear, cherry, raspberry, plum, and so on. But there are many dry fruits, as distinct from the fleshy ones. There are samaras, which are one-seeded fruits with wings (the maple key is a double samara, and the seeds of the ash and elm, single samaras). There are podlike fruits, with seeds inside, that appear in various forms on trees like the locust and the magnolia. But more familiar to us as dry fruits are the nuts, most of which are one-seeded fruits contained in a hard shell. The walnut, hickory, beech, and chestnut all produce this type of seed.

One of the most important fruits of the forest, in terms of what it yields to the woodland community, is the fruit of the oak—the acorn. There are nearly sixty varieties of oak trees in the United States, all producing acorns of one sort or another. What these nuts have in common is their amazing food value, for they are rich in carbohydrates, fat, and vitamins. Most other nuts have to be consumed by gnawing animals because of their hard, protective outer shell; but the acorn, which is seated in a little cup, is one of the staple items in the diet of many insects, animals, and birds.

On the average, oak trees produce bumper crops of acorns once every three or four years, and that means full bellies for all the creatures that feed on them—mice, several types of squirrels, raccoons, deer, and bears. Birds, too, depend on acorns—among them quail, pheasants, wild turkeys, crows, and smaller varieties

like nuthatches and titmice. One of the most unusual birds, which relies entirely upon acorns, is the so-called acorn woodpecker of California, which drills holes in trees and then buries acorns in them, sometimes implanting tens of thousands of nuts in a single tree.

In poor years, when the acorn crop is lean, the situation can be critical for the forest animals. In the summer of 1968, for example, along the eastern seaboard there were almost no acorns, and the squirrels were migrating frantically from one range to another in search of food. They were seen crossing thruways and streams—and the Hudson River—in their desperate quest for acorns. Even in an average year most oaks produce more small, aborted acorns than perfect ones, and only a minute percentage of all the acorns that fall to the ground have a chance of germinating and becoming trees. The average oak produces a crop of about five thousand acorns a year, and in a study made some years ago of a mature tree, scientists learned that 83 per cent of the nuts were eaten by deer and other animals; another 6 per cent had been attacked by weevils and larvae; and most of the remaining acorns were naturally imperfect. They reckoned that less than 1 per cent of the acorns sprouted, and of those that did, half died as seedlings. That particular tree—a prolific one—was found to produce fifteen thousand acorns in the year the study was conducted, but only a handful—between fifty and a hundred—ever germinated successfully. In areas where the deer population is high, surveys indicate that the entire acorn crop may be eaten by the herd; the only years in which any of the seeds have a chance to germinate are when the production of acorns is unusually heavy.

Apparently the acorns from certain oak trees are more appealing to animals than those from other oaks: the red and black oaks produce bitter acorns, while acorns from white oak trees are tastier, so it follows that seeds from the red and black oaks have a better chance of germinating since they are less tempting to animals.

A particularly interesting aspect of the acorn is that it shows, in microcosm, the remarkable interaction between organisms in the process of growth and decay. As the illustrations on these pages demonstrate, the beginning of the end for the acorn often comes while the nut is still on the tree. Wasps or moths may lay their eggs on its surface; the long-snouted acorn weevil bores into it to lay eggs, as do beetles; various forms of larvae or fungi may find their way into the nut—all before it falls to the ground. After it drops to the forest floor, in the late summer or autumn, the acorn may be used in one way or another by a dozen or more forest creatures —some of them microscopic in size, like the nematodes, some larger, like snails, ants, centipedes, and earthworms. When eggs that have been laid in the acorn hatch, the larvae eat their way out, boring holes through the nut. The decaying interior of the acorn is eventually riddled with tiny tunnels; it is partly hollowed out inside; and it becomes a habitation for numerous little animals, as well as for fungi and decay bacteria. When the outer shell of the nut finally collapses, earthworms and other creatures slowly work the remaining debris into the soil.

The reduction of an acorn from nut to humus is shown in these drawings, from left to right. The process of disintegration begins when insects bore into the nut. A squirrel discards the damaged acorn, which is then devoured by wasps or other insects. Ants and a centipede move into the shell, and earthworms work away at the fragments until they are finally reduced to humus.

When early European voyagers approached the land that is now the United States, they were enchanted by the land smell that told them they were nearing shore. This was actually the perfume of a dense, unbroken wilderness that extended from the seacoast to the Mississippi River, and to a man like Arthur Barlowe, who coasted along North Carolina in 1584 and reported his findings to Sir Walter Raleigh, it was a wondrous thing indeed. "The second of July," he wrote, "we found shole water, wher we smelt so sweet, and so strong a smel, as if we had bene in the midst of some delicate garden abounding with all kinde of odoriferous flowers, by which we were assured, that the land could not be farre distant . . ." After going ashore and surveying the rich vegetation, he concluded that "in all the world the like abundance is not to be found: and my selfe having seene those parts of Europe that most abound, find such difference as were incredible to be written. . . ."

It had been hundreds of years since any other European had seen anything even remotely resembling what Barlowe witnessed. There are numerous references to forests in classical writings: Homer speaks frequently of "wooded Samothrace" or the "tall pines and oaks of Sicily," but these woodlands had largely disappeared from the ancient world. The forests had been cleared by man, and his sheep and goats and cattle had altered forever the nature of the landscape bordering parts of the Mediterranean and Aegean. Even at the time of Greece's glory Plato was writing that "What now remains compared with what then existed is like the skeleton of a sick man, all the fat and soft earth having been wasted away, and only the bare framework of the land being left."

To the colonists settling America in the seventeenth and eighteenth centuries the apparently endless woods, unbroken except by rivers and mountain crags, were a wonder that made an enduring impression. It was said that a squirrel could make its way from the eastern end of Pennsylvania to the western boundary without ever leaving the trees, so thick was the cover, and certainly only a bird flying over the ancient forest could take in its immensity. A man traveling through Pennsylvania in 1806 remarked on the "extraordinary height and spreading tops of the trees; which thus prevent the sun from penetrating to the ground, and nourishing inferior

articles of vegetation. In consequence of the above circumstance, one can walk in them with much pleasure, and see an enemy from a considerable distance." Parts of that country were called the "black forest" by pioneers because the vegetation shut out the sun so completely, and General Edward Braddock's route to destiny in 1755 took his army through an almost impenetrable gloom known locally as the "shades of death."

Evidently this rank deciduous and white-pine forest was a world of silence by day, for most of the songbirds that populate more open woods today lived on the perimeters of the dense growth, along the Atlantic coastline and the river valleys and in the few openings that existed in the woodland.

Of an estimated four hundred thousand square miles of virgin forest that once covered the eastern half of the United States, less than two thousand square miles might be said to remain in anything like their primeval state. Only the more inaccessible reaches of the Appalachians and a few small scattered areas elsewhere are left to suggest what the whole region was like. For almost the first two hundred years of American settlement pioneers claimed a farmstead by hacking one out of the forest, cutting down trees so large a man might chop for several days before he could fell one. At best, the pioneer farmer could clear a few acres each year, and one reason so many of the huge trees were girdled was to save a man the backbreaking toil of swinging an axe day after day.

One of the most vivid descriptions of the way immense sections of virgin forest were cleared off was left us by the Marquis de Chastellux, a major general in the French army, who came to America with Rochambeau and kept a journal of his travels for friends at home. Writing in the autumn of 1780, he said: ". . . whatever mountains I have climbed, whatever forests I have traversed, whatever bypaths I have followed, I have never traveled three miles without meeting with a new settlement. . . . The following is the manner of proceeding in these 'improvements,' or 'new settlements,' as they are called. Any man who is able to procure a capital [of about 25 pounds sterling] and who has strength and inclination to work, may go into the woods and purchase a tract of land, usually a hundred and fifty or two hundred acres, which seldom costs him more than a dollar . . . an acre, and only a small part of which he pays in cash. There he takes a cow, some pigs, or a full sow, and two indifferent horses. . . . To these precautions he adds that of having a provision of flour and cider. Provided with this first capital, he begins by felling all the small trees, and some of the big branches of the large ones: these he uses to make 'fences' for the first field he wishes to clear; he next boldly attacks these immense oaks, or pines, which one might take for the ancient lords of the territory he is usurping; he strips them of their bark, or rings them round with his axe. These trees, mortally wounded, find themselves robbed of their honors the following spring; they put forth no more leaves, their branches fall, and their trunks soon become only hideous skeletons. These trunks still seem to brave the efforts of the new settler; but whenever they show the smallest chinks or crevices they are surrounded by fire, and the flames consume what the iron was unable to destroy. But it is enough for the small trees to be felled, and the great ones to lose their sap. This object completed, the ground is 'cleared'; the air and the sun begin to operate upon that earth which is wholly formed of decayed vegetation, that fertile earth which asks but to produce. The grass grows rapidly; there is pasturage for the cattle the very first year; after which they are left to increase, or fresh ones are bought, and they are employed in tilling a piece of ground which, when planted, yields the enormous increase of twenty or thirty fold. The next year, more trees are cut, more 'fences' built, more progress made; then, at the end of two years, the settler has the wherewithal to subsist, and even to send some articles to market; and after four or five years he completes the payment of his land and finds himself a comfortable 'farmer.' Then his dwelling, which at first was no better than a large hut formed by a square of tree trunks, placed one upon another, with the intervals filled by mud, changes into a handsome wooden house, where he contrives more convenient, and certainly much cleaner apartments than those in most of our small [French] towns. This is the work of three weeks or a month; his first habitation, that of twice twenty-four hours. . . . [since] The neighbors, for they are everywhere to be found, make it a point of hospitality to aid the newcomer. . . . Such are the means by which North America, which one hundred years ago was nothing but a vast forest, has been peopled with three million inhabitants."

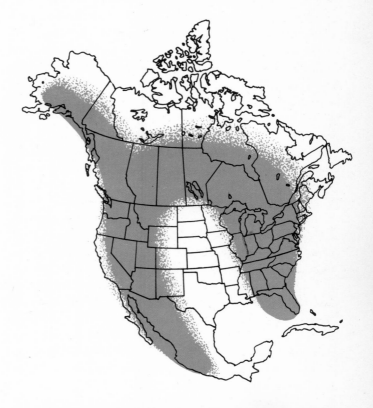

In general, the principal forested lands of North America occupy the horseshoe-shaped arc indicated above in blue.

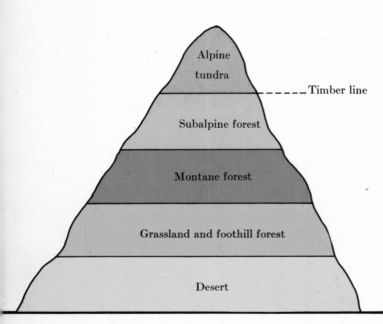

As this drawing suggests, different life zones may prevail on a single mountain, changing from desert conditions at the base to alpine tundra above the timber line.

By the beginning of the nineteenth century frontier farmers reached the western edge of the great eastern forest, coming out into the pocket of the prairie that lies in what is now Illinois. There are records of the joy they felt upon emerging from the gloom and dark of the woods into sunny, open grasslands where they could see the full sweep of sky. The prairie soil was far richer for farming than the land they had left behind, but curiously enough, few of the first settlers knew this. In their experience, and that of earlier generations, something must be wrong with soil in which trees did not grow. And so they settled, at first, in the forest at the edge of the grass. Beyond the Mississippi, in colonial times as now, the woods ended almost entirely and the tall grass of the prairies began—for what reason no one is entirely certain. Some believe that the Indians burned off the forest there to drive out the game and to improve the vegetation that supported it. Others think that the last glacier to slide down over America from the north may have retarded the forest. What is perhaps more likely is that a combination of soil and rainfall conditions was responsible, but no one knows for certain. The early settlers found the prairie grass unusually beautiful, especially when it was interlaced with flowers in the spring, and they wrote that it was so tall a man on foot could not see over it.

Farther west the tall grass gave way to short grass, at a transition line marked by the twenty-inch rainfall boundary running north and south. The terrain here is not rolling; it is almost flat, sloping imperceptibly upward for four hundred miles until it joins the abrupt Rocky Mountains to the west. Nowhere in the country is the rainfall more unpredictable, or the climate more violent. For several years there may be adequate rainfall, then comes a year when there is virtually none. The wind blows constantly; the heat is intense from July through September; and the winters are extreme.

Beyond the Great Plains, the vegetation map is confused, for in this vast area there are forested slopes that catch ample rainfall, dry lowlands and plateaus that produce only dry, thorny shrubs, cacti, and bunch grass, and deserts too dry or salt-laden to support any growth.

The greatest wonder of all are the lush forests of sequoia and fir trees on the western slopes of the mountains, whose peaks catch the heavy Pacific rains. These great trees are among the oldest and largest plants known. In some sections of these woods there is a steady dripping of water onto the vegetation below the trees, because the trees themselves are unable to absorb additional moisture from the humid air. For miles, spruce, hemlock, cedar, pine, and fir climb the western walls of the Cascade Mountains, and farther south, in California, grow the gigantic redwood trees, the monarchs of the North American forest.

Here in the West, this land of extremes, it is possible to see how the different zones of altitude on a mountain slope support a particular kind of vegetation that depends on the conditions of moisture, temperature, soil, and exposure that we have already discussed. The interplay of temperature and rainfall have a marked effect on plants and animals, and plant biologists have identified a number of so-called life zones to describe or classify environments. In the eastern part of the United States, where the zones succeed each other in fairly orderly progression from south to north, the life zones can be overlaid on a map, in proper sequence. In the western mountains, however, we find a number of these life zones existing within a very short distance; by climbing a mountain in Colorado, for instance, it is possible to duplicate ecologically the horizontal distance from the Rio Grande to Hudson's Bay.

At the foot of the mountain, desert conditions prevail. It is a life zone where mesquite and cacti, lizards, rodents, and desert birds are common. The rainfall is well under ten inches, and in certain places it is almost nonexistent. Walking uphill, the hiker passes into a zone where desert grasses and sagebrush begin to merge with piñon pine and juniper—trees ten to thirty feet tall that are sometimes called the "pygmy" conifers. Here there is increased annual precipitation, temperatures that tend to be colder than those of the desert below, and wildlife that includes coyotes, elk, and rattlesnakes.

Still climbing, the visitor moves into the lower reaches of the great montane forest, where ponderosa pines, interspersed with aspens and other deciduous trees, alternate with open, grassy meadows where mule deer graze. Stands of majestic Douglas fir succeed the ponderosa pine at the higher altitudes, and above them—in the highest forest zone—are spruce, fir, and patches of quaking aspen coming up where forest fires have killed off the evergreens. Beyond these trees is the timber line, above which the temperatures are so low, and

the growing season so short, that only the lichens, mosses, and some grasses can support life. These plants extend as far as there is a thin layer of soil or rock on which to secure a foothold; then, at the upper elevations, snow and ice form a perpetual ground cover, and there is no vegetation to be seen.

If we look again at the map on page 55, we see that the forests of North America form the shape of a giant horseshoe, one leg of which begins along the eastern coast of the United States, moves up to arch across the north-central section of Canada into Alaska, and then travels southward down the western extremity of the United States. In the open west-central section few forests exist, because the soil, temperature, and rainfall conditions make it difficult for trees to sustain life. Included in the huge horseshoe-shaped area of forested lands are thousands of different types of vegetation, about which we shall see more on pages 60 to 67, but before examining the principal types of forests that cover North America, we might consider how a forest begins from nothing and goes through a long succession of different stages of growth to become what is known as a climax forest.

For years the New England states were filled with abandoned farms—lands that had been cleared of trees by generations of pioneers determined to scratch out a living from the stony soil. Farmers who were discouraged by increasingly poor yields, or lured by the promise of fertile western lands or the hope of finding gold or silver, packed up their families and belongings, took a last look at the old homestead, shut the door, and departed, leaving behind a few ruinous buildings and fields that had until recently been meadowland and pasture. In the 1870's and well into the 1890's farms were being abandoned wholesale, and one can read between the lines of statistics the profound changes that resulted from this population movement. By 1820 all but 27 per cent of the entire area of Connecticut had been cleared of woods. By 1910 the trees had come back, reclaiming 45 per cent of the land, and by 1955 forests occupied 63 per cent of the state. What happened in the years after New England's farms were abandoned reveals a great deal about the process that is called natural succession. Within a few months of a farmer's departure a long trail of change began to take place—a change that eventually produced a forest to supersede the one cleared long ago at

The symmetrical, fragrant balsam fir, which is so highly valued for Christmas trees, is usually superseded in the eastern forests by one of the climax species. The balsam fir is found over wide areas of the U.S. and Canada.

the cost of such hardship and pain.

Let us imagine that the time is spring, in the year the farm family departed: sunlight falls directly on the soil between the sparse growth in the fields, and by midsummer a collection of new plants has already invaded the territory. Most are weeds—annuals like ragweed, dandelions, and chickweed—whose seeds blow into the fields by the millions or are dropped and spread by the birds. As time passes, other plants seed themselves and begin to grow: goldenrod and Queen Anne's lace, joe-pye weed, black-eyed Susans, milkweed, and mullein. Woody plants like the burdock and blackberry take root, and gradually, patches of soil are shaded from the sun, and more water is held in the earth by the growing number of plants. New animals and insects move into the fields, and an increasing variety of birds populate the area in search of food. Mice, rabbits, moles, and shrews come into the old fields, followed by the predators that feed on them—hawks of all kinds, owls, and snakes. Woodchucks dig their holes and multiple tunnels beneath the soil, and at every level—above and below ground—life proliferates.

Here and there a tree seed has germinated, and a scattering of them—a pine or red cedar, a few white birch and poplar, perhaps a wild cherry—begin to be prominent in the fields. Only those species whose seedlings tolerate direct sunlight move into the open land. In many areas of New England the white pine soon becomes predominant, elsewhere the birch and poplar fill the fields, but none of these species is destined to last permanently. By the time the pines are about fifteen or twenty years old, their branches may form a canopy that darkens the ground, smothering the old pasture grass and weeds, and may begin to form such a complete cover that their own seedlings have insufficient light for germination. And the birches and poplars, which tend to be relatively short-lived, begin to die when they have reached maturity. But these pioneer trees provide shade for still other species that are far more tolerant of it—trees that actually require a certain amount of protection from direct sunlight. Red oaks and maples, tulip poplars and ash, achieve a foothold under the pines or birches, infiltrating the woodland slowly but persistently. In the pinewoods the weeds

One of the pioneer trees is the eastern red cedar (which is not a true cedar, but a juniper). It grows from the Atlantic coast to central Kansas, often in very poor soil. Under ideal conditions it may reach 100 feet in height.

disappear as they are deprived of sunlight, and with their departure the character of animal life alters, too. Little animals that made nests of grass in the fields move out, deer move in, and the field birds are supplanted by those that live in or on the edges of forests. Chipmunks and squirrels are busy carrying acorns, butternuts, or hickory nuts into the woods, and some of these take root with the passing of time.

Inevitably, the pines begin to disappear one by one—some of them blown over by windstorms, some destroyed by fire or lightning, still others damaged by insects or disease—and as they topple over, their places are taken by trees that have been growing patiently on the forest floor—usually red maples, red oaks, or tulip trees. Similarly, the white birch is replaced by other types of trees: the line of succession is from the sun-loving pioneer trees to those that can grow in partial shade, and these latter form a forest that is still only in the intermediate stage. For, like the pines, these trees eventually bring about their own destruction by growing so large that their own seedlings are incapable of surviving in their shade. During the passing decades the species that

thrive in deep shade—sugar maples, hemlocks, beeches, white oaks, and hickories—have been taking over the forest floor from the other plants, forming a dense understory, and each time one of the red oaks or red maples succumbs to old age or some catastrophe, they are released and grow more quickly, filling in the space until the forest finally achieves what is called its climax stage. From this time on, unless a major calamity befalls the forest, only those seedlings that can grow in dense shade will survive. And below the understory the cycle of growth and decay, the building of humus and topsoil, is going on steadily, adding riches to the earth and creating the conditions of moisture and food that nurture the big climax trees.

From studies made of true wilderness tracts, it is estimated that a climax forest of this type—a balanced community of plant and animal life—takes between two and three hundred years to replicate. When woodland such as this is destroyed by man or by natural catastrophe, the entire cycle of succession must begin again, and it is a matter of centuries—not years—before the climax forest can re-create itself.

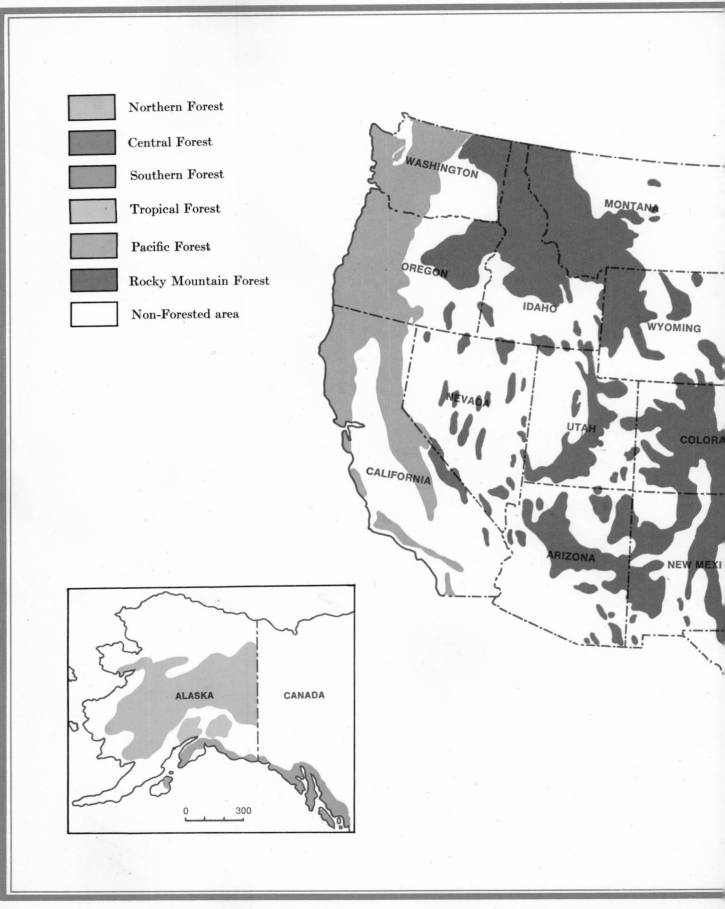

Northern Forest

Central Forest

Southern Forest

Tropical Forest

Pacific Forest

Rocky Mountain Forest

Non-Forested area

FOREST ZONES of the U.S.

NORTH DAKOTA

MINNESOTA

SOUTH DAKOTA

WISCONSIN

MAINE

VT.

N.H.

NEW YORK

MASS.

R.I.

CONN.

MICHIGAN

20 inch rainfall line

NEBRASKA

IOWA

PENN.

N.J.

OHIO

MD

DEL.

ILLINOIS

INDIANA

WEST

VA.

VA.

KANSAS

MISSOURI

KENTUCKY

NORTH CAROLINA

TENNESSEE

SOUTH

CAROLINA

OKLAHOMA

ARKANSAS

MISS.

ALA.

GEORGIA

TEXAS

LOUISIANA

FLA.

0   400

scale of miles

The West Coast Forest is characterized by tall, stately conifers like the Douglas fir and—most notably—the redwoods, shown in this illustration, whose growth is favored by the abundant moisture continuously blown in from the Pacific.

Within the great horseshoe-shaped arc of trees that covers such vast areas of the North American continent, there are six principal forest regions—about equally spread geographically between east and west. Because of its great size, and because of the varied conditions of climate and soil that prevail in different sections of the continent, North America is favored by a particularly rich variety of vegetation; it has, in fact, approximately four times as many different tree species as Europe does.

Along a thin strip of land running for three thousand miles, from Kodiak Island in Alaska to the Santa Cruz Mountains near San Francisco, is the Pacific Forest—a prodigious growth of conifers unequalled anywhere on earth, which has been created by a combination of temperature, rainfall, and the topography of the region. When the warm, moisture-laden air blown in by the westerly winds strikes the Pacific coastline, it hits the high mountain barrier almost immediately, rises, is cooled rapidly, and condenses into snow or rain. The coast ranges rise to a height of 14,000 feet, and by trapping enormous quantities of moisture, they make possible the existence of rain forests in portions of this temperate zone.

These conditions have produced the largest living organisms—the giant sequoias and the coast redwoods, found in northern California. The sequoias climb the slopes of the Sierra Nevada range, at altitudes of 4,500 to 8,000 feet, while the redwoods are concentrated in an area about 500 miles long and 15 miles wide in foggy areas near the ocean. Some of these gigantic specimens are more than 3,000 years old, with a mass of roots 3 or 4 acres in size; the redwoods reach towering heights of 300 feet and more, and giant sequoias are known to

The Mountain Forest of the West has less rainfall than the West Coast Forest because moisture from the Pacific is blocked by the Cascade and Coast ranges. A prominent species of trees is the ponderosa pine, in the foreground.

have trunks measuring up to 30 feet in diameter.

. Farther up the coast is the Olympic rain forest of Washington, with its dense stands of western hemlock, western red cedar, Sitka spruce, and Pacific silver and grand firs. Here, in this lush forest, water particles literally saturate the air and drip continuously to the floor, making it as luxuriant as a tropical jungle, with mosses so thick and moist that the visitor sinks into them up to his ankle. Douglas fir, which is the nation's most important commercial tree (after more than three decades of lumbering it still accounts for one-fourth of the total volume of timber standing in the United States), is concentrated in Washington and Oregon, and northward are endless reaches of spruce skirting the glacial lakes and bogs of Canada. In Alaska, western hemlock and Sitka spruce are the predominant species; between them, they cover nearly 4-1/3 million acres.

To the east of the Pacific Forest, in the rain shadow of the coastal ranges, lies the immense area of mountain and desert including the states of Arizona, Colorado, Nevada, New Mexico, and Utah in the south, and eastern Washington and Oregon, Idaho, Montana, Wyoming, and the western portions of South Dakota at the northern end. Because it receives so much less rain, this Rocky Mountain Forest is more sparsely covered with trees than the section along the coast, but it is nevertheless heavily forested. The most important trees found here are ponderosa pine, western white pine, Douglas fir, lodgepole pine, and Engelmann's spruce. It is said that Spanish ranchers used to burn off the mountain forests to create pastureland for their goats and sheep, and American prospectors did the same thing— it was the easiest way to get rid of the trees that hindered their search for outcroppings of precious metals.

The Northern Forest spans a region of many lakes, rivers, and hills. Some areas have mixed species, as this drawing shows; others are predominantly coniferous (pine, hemlock, spruce); still others, hardwood (maple, birch, beech).

These and later fires made it possible for the big stands of lodgepole pine to take hold, for the seeds of this tree germinate by the millions in ash-covered soil, covering the fire scar with a thick blanket of seedlings. Here, in the foothills of the Rockies, the "pygmy" conifers grow, and piñon pine and juniper cover millions of acres. In the subalpine zone of the mountains spruce and fir predominate, and they are succeeded near the timber line by larch, whitebark pine, limber pine, and bristlecone pine. The last is the tree that has been discovered to be the oldest living thing: clinging to life along the arid heights of the White Mountains of California are trees found by scientific investigation to be more than four thousand years old.

A third forest area forms the arch of the horseshoe-shaped range of trees and spans the eastern and western groups of vegetation. Spreading some four thousand miles across the continent, from Alaska to Newfoundland and down the higher peaks of the Appalachian chain, is the Northern Forest. Above it is tundra, thin-soiled land of low shrubs and lichens, which disappears into the Arctic icecap—a reminder of the vast glaciers that scraped off the hilltops of Canada and left so much of it flat, with poor drainage, dotted with lakes and bogs known as muskegs. In the north woods white and black spruce grow, the latter being particularly tolerant of the cold, wet, windy climate, and there are tamarack, aspen, and paper birch. The soil throughout the Northern Forest expanse is generally poor, and five of the most characteristic trees are conifers—balsam fir, eastern hemlock, northern white cedar, and eastern white pine. The most prevalent hardwood is yellow birch; while it is true that the sugar maple invades the Northern Forest, that species is somewhat more typical

 The Central Forest is typified by hardwood species. A great deal of the region was converted to farmland in the past, but it contains various important species such as oak, hickory, yellow poplar, yellow pine, and red cedar.

of the southernmost regions.

This was an area that supplied much of the lumber used in the first 250 years of settlement in the United States—most of it eastern white pine—and as the virgin forests of the east were gradually depleted, the loggers moved westward. In the mid-nineteenth century Maine, New York, and Pennsylvania were the leading timber producers; by the third quarter of the century Michigan, Minnesota, and Wisconsin had replaced them. With the wholesale logging of the great white and red pine forests, and disastrous fires that accompanied it, the humus beneath these wooded lands was destroyed, and along with it the seedbeds of the white and red pine. In their stead the jack pine, now so typical of the states bordering the Great Lakes, grew up, to become a dominant type in certain sections. When the cones of the jack pine are heated, they scatter huge

numbers of seeds on the ground; a study of Minnesota fires indicated that where there might have been only a handful of jack pines in an acre of woodland prior to a fire, as many as twenty thousand seedlings per acre could spring from the ashes afterward.

Other species that inhabit the Northern Forest are beech, aspen, walnut, hickory, ash, and basswood in the northerly region; yellow poplar, black tupelo, and many varieties of pine and oak in the south. The most abundant tree to have flourished in the Appalachians is now virtually extinct: the American chestnut was wiped out by a blight that was carried into this country on some Asian plants early in the twentieth century.

Spreading out from the Northern Forest into the lower elevations of New England and New York, on into Ohio and down to Arkansas and Oklahoma, is approximately 130 million acres of land whose forest cover

The Southern Forest is dominated by southern yellow pines, especially loblolly, slash, and longleaf pines, shown here. But it is also a region of hardwoods—among them sweet gums, tupelos, and a particularly large number of oaks.

has enough common characteristics to enable us to group it under one collective heading. This is the so-called Central Forest, which touches thirty states from Cape Cod to the Rio Grande and back up to the Canadian border. The trees that occupy it are mostly deciduous, which means that it is for the most part green in summer, bare in winter, and a brilliant display of foliage in the autumn, just before the leaves fall to the ground. These hardwood trees grow on soil that was pushed south in the ice age, and many of them have paid the penalty of affluence: the soil was too rich to be left to trees, and for generations they have been cut to make way for farmland. The sugar maple and beech are common to this forest zone in New England; in Pennsylvania and the Middle West they give way to the oaks and hickories; and in the southern states sycamore, tulip poplar, sweet gum, and several types of pine are characteristic. Although more than forty different species inhabit the Central Forest, only one—the white oak—is found throughout the area.

The fifth major family of trees—the Southern Forest—occupies the coastal plains of the Atlantic from southern Virginia deep into Florida and extends through the Gulf States out into Texas and Missouri. If this land area is compared with a map showing the geographic distribution of cotton, the two may be seen to be quite similar: indeed, when the worked-out fields of cotton were abandoned, the sandy soil often reverted to pine, which is now so typical of the Southern Forest. There is a rich variety of trees in the Southeast—at least half of all the species in the country—and they include numerous hardwoods: mostly oaks, baldcypress, red maple, elm, and cottonwood, which line the swamps and bottomlands. But the dominant trees are

The Tropical Forests of the United States are found in two restricted areas—the southern tip of Florida and a small section of southeastern Texas. Mangroves (in the foreground) and mahogany are the principal trees of the region.

four important pines—the shortleaf, longleaf, slash, and loblolly. At the time the white and red pine stands of the Great Lakes states were nearly exhausted, about 1900, the Southern Forest took the lead in timber production and held it for a decade, until it was supplanted by the woodlands of the Pacific Northwest; but this region's relatively high rainfall and long growing season continue to produce trees in great abundance, which have been found more suitable for high-grade wood pulp than for lumber. At one time this area was responsible for almost two-thirds of the world's production of turpentine and rosin, and today the crude resin and gum from its pine trees still furnish the raw materials of a naval-stores industry.

At the very tip of Florida, and in a tiny area of southeastern Texas, is the smallest forest zone of North America. It is a land where winter never comes, where the temperature variations are minor, and where the only perceptible difference between seasons is in the amount of rain that falls. This is the Tropical Forest, typified by the Florida Everglades, whose remarkable mangrove trees form some of the most impenetrable growth on earth. Thriving in salty sand, the mangrove's arching roots rise from the salt water after finding a foothold, and begin to create land where only water existed. The roots burrow into the sand and form a tangled mass of debris that forms islands of soil. Most of these trees never attain great size or height, but some do grow to be sixty feet tall. Neither the mangroves nor the other trees in this region—mostly broadleaf species like palms—are significant commercially, but they are North America's closest approximation to the exotic vegetation that prevails in tropical areas nearer the equatorial zone.

272'

221'

162'

160'

122'

116'

102'

96'

78'

Giant Sequoia
Sequoia Natl. Park, Calif.

Douglas Fir
Olympic Natl. Park, Wash.

Ponderosa Pine
Lapine, Ore.

White Birch
Lake Leelanau, Mich.

American Elm
Trigonia, Tenn.

When, in 1953, scientists studying the age of the bristlecone pine discovered that seventeen of these trees in a California grove were 4,000 years old and that one was 4,600 years old, they realized that this sparsely needled tree was the world's most venerable living thing. What was more astonishing than the fact of such incredible longevity was that it had been achieved in one of the most rigorous environments imaginable—for the bristlecones grow at altitudes between 8,000 and 11,000 feet in the Rocky Mountains of Colorado and Utah and in remote corners of Nevada and eastern California—high in the subalpine forest where they must survive the hazards of poor soil, inadequate moisture, and piercing winds. Size, of course, is not necessarily a direct function of age, for the bristle-

cone pine never attains great stature, any more than the lichens do, for all their long, patient growth.

On the other hand, the largest trees in the United States *are* very nearly the oldest. As Donald Culross Peattie writes, "Of all that has survived from the Mesozoic, which began two hundred million years ago and ended about 55,000,000 B.C., Sequoia is the king. It is so much a king that, deposed today from all but two corners of its empire, superseded, outmoded, exiled and all but exterminated, it still stands without rival. And from all over the world, those who can make the pilgrimage come sooner or later to its feet, and do it homage." The sequoia, of course, is represented in North America by two species—the coastal redwood and the giant sequoia, known as the Sierra redwood or simply

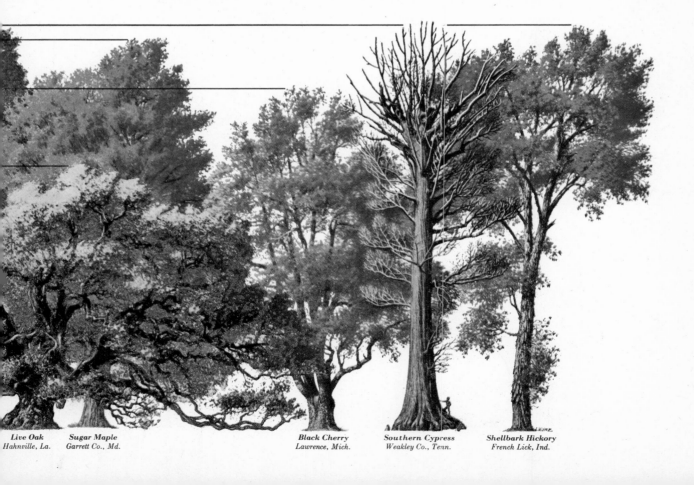

Live Oak
Hahnville, La.

Sugar Maple
Garrett Co., Md.

Black Cherry
Lawrence, Mich.

Southern Cypress
Weakley Co., Tenn.

Shellbark Hickory
French Lick, Ind.

as the big tree, which John Muir called "king of all the conifers of the world, 'the noblest of a noble race.'" Founders Tree, a coastal redwood in Humboldt State Park in California, was measured in 1947 and found to be 364 feet tall, and from the study of annual rings in similar trees, we know that these forest patriarchs occasionally live for more than two thousand years. A giant sequoia known as the General Sherman is 273 feet tall and 115 feet in circumference; it is one of a number of such specimens whose life span reaches or exceeds three milleniums, and it may have been living as long ago as the fall of Troy.

In tropical rain forests, where the canopy may be 150 to 200 feet above ground, some immense trees tower 260 feet in the air; their life expectancy seems to be from three to four hundred years. For some physiological reason that is not entirely understood, most small trees and shrubs tend to have shorter lives than large ones.

A plant must, of course, continue to grow if it is to survive, and this is one of the fundamental differences between plant and animal. The latter, once it attains full size, may live for many years without adding to its stature; the reason is that the animal's cells are capable of longevity, even though they may not be capable of replacement. The tree, on the other hand, has cells whose durability is much more limited, and it must replace them in order to sustain life and grow. Because of this, a good many trees are longer-lived than man is, but there is nevertheless an outer range of longevity and size for each type of tree.

The Mullan Tree (far left), a western white pine in Coeur d'Alene National Forest, Idaho, was named for John Mullan. It marked a road built under his direction from Fort Benton, Mont., to Walla Walla, Wash.

Until it blew down in 1856, the Charter Oak (center) was the most famous tree in United States history. This great oak provided a hiding place for the Connecticut charter when King James II demanded its surrender.

Indian-trail trees (right) often marked a trail for hunting or trading. Indians bent young saplings over and tied their tips to the ground. Trees like this one, near Evanston, Illinois, sometimes took root a second time.

A number of pines found in the Northern and Southern forests—among them the white, loblolly, and short-leaf—grow to heights of a hundred feet or more at maturity; by contrast, the jack pine is commonly only twenty to thirty feet tall. One of the largest and most beautiful trees in the eastern United States is the graceful, fountainlike American elm, which typically grows to a height of eighty or a hundred feet, while larger specimens—like the one illustrated on page 68—are found from time to time.

Men have always been fond of recording examples of great size or span or age in trees; Alexander the Great is known to have marveled at the dimensions of an Indian banyan tree (although it was not the same tree, there is a record of a similar banyan in the Andhra val-

ley that had a crown measuring two thousand feet in circumference, supported by 320 trunks). When the North American continent was newly settled, there were highly practical reasons for noting distinctive trees or those of great size: they served as landmarks in an unbroken wilderness, indicating the route of a trail or marking off the distance traveled by a man on foot. And old deeds are sprinkled with references to a giant oak or beech or hemlock, from which an early surveyor ran his chain to another boundary line. (A magnificent white oak, still standing in 1970 near Abraham Lincoln's Kentucky birthplace, was a landmark when the Civil War President was born in 1809; even then, it was known as a "boundary oak.") Prior to the Revolution, tall, straight white pines with trunks more than two feet

in diameter at the base were reserved for masts for the Royal Navy, and Royal Tree Viewers blazed them with the king's broad arrow to indicate that they were crown property.

Men's taste for the distinctive often led them to associate a particular tree with an important man or event, so many assumed the status of memorials as well as landmarks. Nearly every town in the colonies, both before and during the Revolution, had a liberty tree. A great elm in Boston, upon which unpopular British ministers were hanged in effigy in the furor that followed the Stamp Act, was the town's Liberty Tree until it was cut down by British soldiers during the occupation. After the Revolution the countryside proliferated with trees supposedly associated with George Washing-

ton. One of the best known was the "Washington Elm" in Cambridge, Massachusetts, beneath which he reputedly assumed command of the army outside Boston in 1775. Another elm, under which the first President supposedly stood while he watched the construction of the new Capitol, stood until 1948. And there are today at Mount Vernon numerous trees planted during Washington's lifetime—a variety of tulip trees, hollies, buckeyes, elms, hemlocks, pecans, and mulberries. Jefferson shared Washington's love for plants of all kinds, and his journals are filled with references to them. He imported plants from all over the world, keeping careful records in his *Garden Book* of the growth of his orchards.

The huge Charter Oak in Hartford, Connecticut, was a historic shrine for nearly two hundred years. In 1687,

The Sailor's Sycamore (far left) was used by navigators as early as 1800 to locate their anchorage as they sailed into port. The tree stood on what is now the corner of Milpas and Quinientos streets in Santa Barbara, Cal.

Until 1917 a beech known as Daniel Boone's "Bar" Tree stood in northeastern Tennessee. Here the great explorer of the Appalachian Mountains carved the words: "D. Boon cilled A BAR On Tree in THE YEar 1760."

The so-called Dueling Oak—a live oak that stood in New Orleans' City Park—dated back to the early Creole days of Louisiana, when real or fancied slights were often an excuse for an affair of honor "under the oaks."

when Sir Edmund Andros demanded the surrender of the colony's charter, a local patriot supposedly hid the document inside the oak. This important piece of paper served Connecticut as a constitution from 1662 until 1816, granting it all lands "from the said Narragansett Bay on the east to the South Sea [*i.e.*, the Pacific] on the west." By the time the oak blew down in 1856, the hole in which the charter had been concealed had increased in size sufficiently to hold twenty-five men, it was said.

According to local tradition, a group of majestic hemlocks in Germantown, Pennsylvania, were set out by William Penn; a venerable oak in Crockett, Texas, was said to mark one of Davy Crockett's camp sites when he was en route to Texas and the fight at the Alamo;

and the Evangeline Oak at St. Martinville, Louisiana, stood there when the Acadians landed in 1758, after they had been driven from Nova Scotia. Until recent years Indian-trail trees could be seen in the Mississippi Valley—trees that were bent over when they were saplings to mark the direction of a trail. As the illustration on page 71 indicates, they sometimes took root from the point at which they touched the ground.

Several live oaks that were standing in 1814, just behind the line held by Andrew Jackson's troops at the battle of New Orleans, survive in Chalmette National Historical Park. And at least seven oak trees are left from the copse on Cemetery Ridge near Gettysburg, which marked the target of Pickett's Charge against the Union line on July 3, 1863.

Writing from Philadelphia in 1793, Thomas Jefferson remarked in a letter to his daughter, "I never knew before the full value of trees. My house is embosomed in high plane trees, with good grass below, and under them I eat breakfast, dine, write, read and receive my company." There is no counting all the other Americans who have lived in the eastern half of the United States who have derived similar spiritual satisfaction from its infinitely varied broadleaf forest—a forest that also yields shade, shelter, and fuel and serves as a bountiful source of food, clothing, tools, vehicles, housing, and a wondrous number of other products.

Known variously as deciduous, or broadleaf, or hardwood, the trees that dominate this region are the culmination of the flowering plants that came into existence some 300 million years ago. The word "deciduous" indicates that they shed their leaves in the fall to seal in the tree during winter when its source of moisture is cut off (in warm climates the leaves of some deciduous species remain on the tree all year, but are replaced annually). "Broadleaf" is sometimes used to describe them because of the foliage that is typical of most such trees. And "hardwood" is the term given them by lumbermen, for the character of timber produced by many of them (it is by no means a fully accurate description, since some softwoods are actually harder than some of the trees that are classed as hardwoods).

The shape of the shagbark hickory shown on the opposite page in both its summer and winter guise is fairly representative of deciduous trees that grow in the open, unhindered by competitors in the forest. That is, they have a straight trunk that sometimes divides into two or more parts, from which many spreading branches grow outward and usually upward, producing a broad, somewhat rounded form. Unlike the conifers, the deciduous trees have no terminal bud at the top that accomplishes all upward growth.

These broad-leaved trees—of which there are about 650 species in the United States—belong to the large group of plants known as angiosperms that also includes most species of garden and wild flowers. The term indicates that their seeds are encased in a protective body; like the conifers, they bear seeds, but they have developed a variety of flowers to attract insects and animals for pollination.

One difference between the needle-leaved trees and the broad-leaved is that the latter have special tubes or vessels in the wood structure that carry sap; experiments show that water travels at a rate up to seventy times as fast in oak and ash trees as in the conifers—a difference that handicaps the conifers when they are competing with broad-leaved trees in the warm, moist climate conducive to rapid growth.

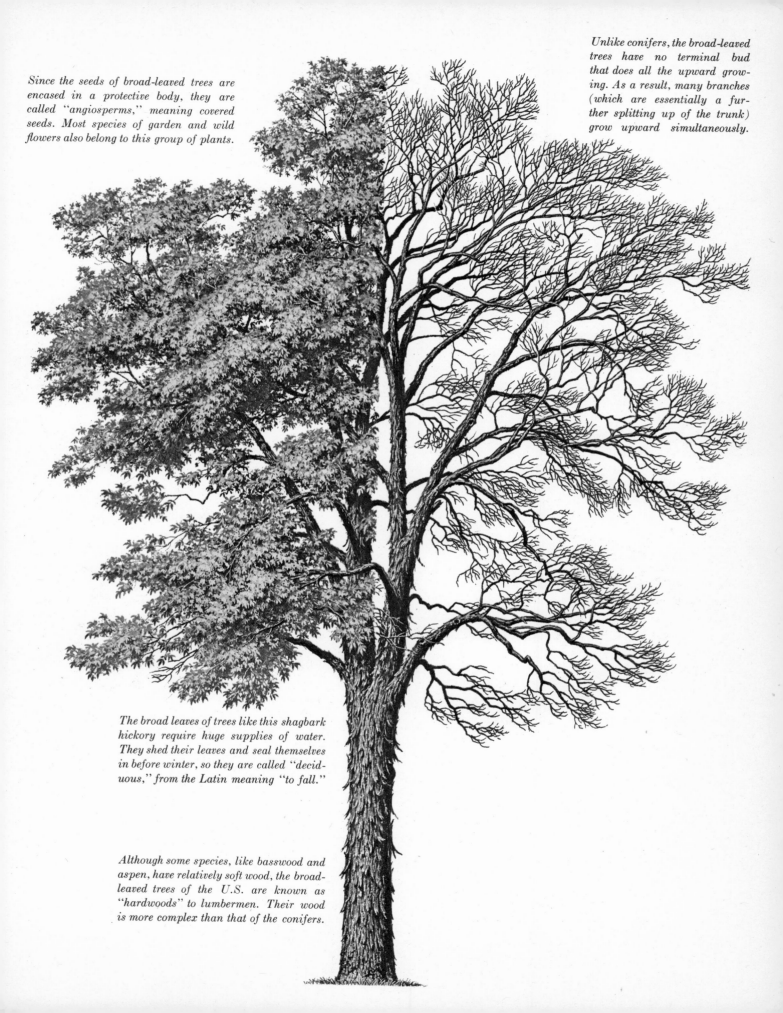

Since the seeds of broad-leaved trees are encased in a protective body, they are called "angiosperms," meaning covered seeds. Most species of garden and wild flowers also belong to this group of plants.

Unlike conifers, the broad-leaved trees have no terminal bud that does all the upward growing. As a result, many branches (which are essentially a further splitting up of the trunk) grow upward simultaneously.

The broad leaves of trees like this shagbark hickory require huge supplies of water. They shed their leaves and seal themselves in before winter, so they are called "deciduous," from the Latin meaning "to fall."

Although some species, like basswood and aspen, have relatively soft wood, the broad-leaved trees of the U.S. are known as "hardwoods" to lumbermen. Their wood is more complex than that of the conifers.

The colors on the map and the matching dots next to each tree show the principal areas where it usually grows.

White oak

Sweet gum

Shagbark hickory

Yellow poplar

Sugar maple

In a typical deciduous forest in the Central Zone the soil and climate produce a profuse variety of vegetation, and the slowly decomposing leaves that fall in the autumn are transformed into a fertile humus for plant growth of all kinds. Because many species of trees and other plants require similar soil and climatic conditions, they are often found growing side by side in certain regions, and the names of those species that dominate a particular community are used for identification purposes. Where they occur repeatedly together, they are known as "associations." In other words, broadleaf angiosperms of all kinds occur throughout the extensive Central Forest, but in West Virginia we may find that a tulip tree-oak association is typical while in Minnesota a maple-basswood association is more common. Throughout the eastern United States there are a variety of these groupings: beech-maple, oak-hickory, hemlock-white pine-northern hardwood, and others.

In whatever region they have been predominant, those were the trees that built and furnished America's homes and provided the lumber so essential to the country's needs. Among them are certain individual hardwood species, illustrated on these pages, which continue to be some of our most important and useful trees.

White oak grows throughout the eastern half of the United States, from Maine to Florida and as far west as Minnesota and Texas. The biggest specimens are found in rich, well-drained soil, particularly on the western slopes of the Allegheny Mountains and in the Ohio River basin. Although it is commonly about 80 to 100 feet in height, individual trees, centuries old, are known to have been 150 feet tall with diameters of up to 8 feet. The strength of the wood, its tight growth-rings, and its durability made it a favorite for a variety of uses, including construction of bridges, barns, ships, wagons, and railroad ties, and it was the lumber always employed for kegs, barrels, and buckets. Today it is used extensively for flooring, furniture, and cabinetwork.

Sweet gum is a member of the witch-hazel family, and it flourishes from southern Connecticut to Florida and west into Texas and southern Missouri. One of its distinctive features is the star-shaped leaf, which gives off a balsamlike fragrance when it is crushed. In the bottomlands of the Southeast, where it often grows in association with red maple, elm, ash, cottonwood, and oak, it grows as high as 80 or 120 feet, with trunks up to 3 feet thick. A strong, stiff wood, it is often used for furni-

ture, cigar and other boxes, and barrels, and it is one of the important sources of plywood.

Shagbark hickory, whose name comes from the distinctive long plates of gray bark on the mature tree that are loose at one or both ends, is found in almost all eastern states from Maine to Minnesota and from the Canadian border to southern Alabama. The sweet-kerneled nuts, which are such favorites with squirrels, were made into a liquor by the Indians; American farm families often used the nuts in candies and cakes and the wood for smoking bacon and hams. The hickory has the toughest, strongest, and most elastic wood of any commercial hardwood, which has made it the principal source of handles for many tools, and for years it was the material from which wheel rims, spokes, and shafts for buggies were fashioned. Other uses today include athletic equipment, such as archery bows and gymnasium apparatus, and furniture of many kinds.

Yellow poplar, also known as the tulip tree, is one of the biggest and most valuable trees found in the southeastern United States. Millions of years ago other members of the same family grew in the Northern Hemisphere, but since the ice age only two species remain—the American tree and one that grows in China. Found from southern New England to Florida and west to the Mississippi River, the yellow poplar often attains great heights, and not infrequently a giant tree will be seen without a branch for the first eighty or one hundred feet. Despite its name, it is not actually a member of the poplar family; it belongs to the magnolia group and is only labeled a poplar because of its soft yellow wood, which is used in the making of crates and boxes and in the core of some plywood.

Sugar maple, the source of maple syrup and sugar, is one of the great ornamental and timber trees of the Central Forest. Its range is from New Brunswick west to Minnesota and south into Tennessee and Arkansas, and because of the brilliant yellows and oranges of its leaves in the fall it is one of the chief contributors to the rich tapestry of autumn color in the Appalachian Mountains. From colonial days it has been prized as a wood for furniture making, and the tiger, or curly, and bird's-eye grains that are occasionally found in these trees have been sought after by gunsmiths and cabinetmakers for many years. Between thirty and forty gallons of its sap are needed to boil down one gallon of syrup—a process the early settlers learned from the Indians.

In the autumn, as if by some secret signal, the leaves of deciduous trees begin to change color, from their customary lush green to a riot of fiery hues—brilliant yellow and orange for the sugar maples, crimson in the soft maple, the delicate yellow of the paper birch, purple for ash, and the bright red of sumac. Because it has such a diversity of species, the North American broadleaf forest presents a spectacle in the autumn of the year that can be seen nowhere else.

Frost does not effect this change; the internal workings of the leaf are affected by the onset of cold, drier weather, and in preparation for sealing off the tree's supply of water for the winter, a band of cells called the abscission layer forms at the base of each stem, where the leaf is attached to the twig. Beneath this layer another one forms, like scar tissue, to cover the wound when the leaf falls, and this growth stops up the pipelines of the tree so that no more moisture goes into the leaf for a period of about two weeks. This is the time of the colorful fall foliage, which occurs when photosynthesis stops. Since chlorophyll must be renewed in order to survive, it disappears from the leaves when the

flow of moisture ceases; and once the dominant green color is removed, the leaf reveals other hues that were there all the time, obscured by the green.

Each tree has its own special chemical make-up that provides its distinctive coloration. Sugar maples and birches are rich in carotene—the pigment found in daffodils, carrots, corn, and butter—which causes them to turn bright yellow or orange. The scarlets and purples are created by a chemical called anthocyanin, which functions in a manner similar to litmus paper: if the sap in which it is dissolved is acid, it turns red; if the sap is alkaline, it changes to blue or purple. This is the same pigment found in cranberries and Concord grapes, and is the one that turns the acidic soft-maple leaf red and the alkaline ash to purple. Brown pigments known as tannins give the oak leaves their rich brown color.

The leaves of some trees are not particularly colorful in the autumn; they fall to the ground without changing their shade when the brittle abscission layer forms, joining others whose colorful pigmentation gradually disappears until the forest floor is carpeted with a thick brown layer.

The green color of the leaf comes from chlorophyll. When it is deprived of its fresh water supply, it breaks down and disappears.

Then yellow and orange pigments, called carotenes, formerly hidden by the chlorophyll, are visible.

Reds and purples can be seen in some hardwood species in autumn.

At summer's end an abscission layer forms where the leaf joins the twig, and shuts off the water supply.

When that alert traveler the Marquis de Chastellux was making his way through Connecticut in 1780, his instinct for accuracy was offended by the way the natives referred to pines, cypresses, firs, and other conifers indiscriminately as "pine trees." It is a habit that persisted, largely because so many species have the typical tall, shaftlike trunk that extends to the very top of the tree, giving them a pyramidal form so familiar in our traditional Christmas trees. Indeed, botanists used to classify all conifers as members of the pine family, but recently the practice has been to divide them into four groups: the Pine family, which includes pines, spruces, firs, hemlocks, and larches; the Baldcypress and Redwood family, represented by the baldcypress and the sequoias; the Cypress or Cedar family, including true cypresses, arborvitaes, incense and white cedars, and junipers; and the Yew family, represented by yews and torreyas.

Whatever group they belong to, the trees are generally known by several different terms: "conifer," because they bear cones; "evergreen," because they customarily retain their needlelike leaves through the winter (a few, like the larch, do not); and "softwood"— the lumbermen's term—because their wood is often lighter and softer than that of hardwoods.

The earliest seed plants to survive until modern times are the conifers, and their remarkable adaptability has left them virtually unchanged for more than 300 million years. The cone that is responsible for their success is essentially a tight concentration of spore-bearing "leaves," or scales. After fertilization the spores produce seeds at the base of the scales, and the cone structure protects them until it ripens, when the scales open and the seeds can emerge. Since these seeds are not encased in a protective body, the trees are known as gymnosperms, meaning naked seeds.

The needlelike leaves of such trees as the longleaf pine shown here are tough, leathery, and covered with a weatherproof waxy skin. These narrow leaves are usually replaced after two years (new needles grow annually on most conifers), but the bristlecone pine is one species that clings to its needles—for as long as fifteen years or more. The trees' typical growth pattern is simple and symmetrical, with branches radiating from the straight central stem almost at right angles, like spokes of a wheel. As a rule, the vertical distance between the whorls of branches normally represents a year's growth. A single terminal bud, or leader, at the top is responsible for the upward development that occurs during each year of the tree's life.

Narrow-leaved trees are known as "conifers" because their seeds are carried in cones. Since the seeds are not encased in a protective body, the trees are also called "gymnosperms," which means naked seeds.

At the top of the simple, symmetrical narrow-leaved trees is a single terminal bud, which is responsible for all the upward growth and adds to the height of the tree during each successive year of life.

The needlelike leaves can live through the winter, giving rise to the term "evergreen" to describe these trees. Needles are covered with a weatherproof wax and are generally quite tough.

The wood of narrow-leaved trees usually has a simple cell-structure and is often softer and lighter than that of broad-leaved trees. As a result, lumbermen refer to the narrow-leaved trees as "softwoods."

J. KUNZ

In much of the Northern Forest, where firs and spruces predominate, the soil is entirely different from that found in the deciduous woods. As the needles slowly decay, the litter collects in a thick humus known as *mor*—a highly acidic blanket from which the falling rains and snow move acids deep into the soil, where they destroy many subsurface minerals. This high acidity and low fertility, in combination with the heavy mat of needles on the ground and the subdued light that penetrates heavy stands of evergreens, cuts down the number and variety of plants on the forest floor and makes it a less appealing territory for small animals and soil organisms. On the Pacific Coast, however, and in certain areas of the Rocky Mountains, there is less acidity because the needles decompose more quickly in the milder climate, and here the soil is comparable to that of the forests in the Central Forest region.

As with the hardwoods, certain species of softwood trees are particularly significant as rich sources of wood for man's use. The seven trees shown here are examples of the conifers that make up nearly four-fifths of our standing timber.

Eastern white pine, largest of the northeastern conifers, was the foremost timber tree of colonial America—the choicest of which were marked with the king's broad arrow, reserving them for use as masts for the Royal Navy. It is estimated that the original stand in North America amounted to 750 billion board feet of timber, but after two centuries of heavy lumbering the virgin forests of eastern white pine were all but gone by 1900. The wood is easy to work, which made it a favorite for the construction of homes and barns, furniture, and many other products, including matchsticks. The tree's range is from Newfoundland to Manitoba and down the Appalachian chain into northern Georgia.

Douglas fir is considered to be the most important timber tree in the world. In ranges along the Pacific Coast and through the Rocky Mountains, and on the Pacific slopes it reaches magnificent proportions: trees more than 300 feet tall have been recorded, with trunk diameters of up to 17 feet. In terms of its weight it is the strongest American wood, yet it is easy to work, and its great size makes possible the cutting of lumber remarkably free from knots. Its principal use is in building construction.

Longleaf pine, one of the largest of the southern pines, was once comparable to the Douglas fir in importance as a timber tree. Its strong, hard wood is highly desirable for construction, and it is also used for ship-masts, railroad ties, and flooring. Cut for pulpwood, it is made into paper cartons and bags and is one of the principal trees used for the production of naval stores—paint, varnish, turpentine, and rosin. It grows in the nearly subtropical climate of the southern states bordering the Atlantic Ocean and the Gulf of Mexico and frequently reaches heights of 100 to 120 feet and a diameter of 2 to 3 feet.

Western hemlock thrives in the dark forests of the humid Pacific Northwest and ranges from northern California to Alaska and inland as far as Montana. In trees as old as five hundred years heights of up to 250 feet have been recorded, and great trunks up to 10 feet thick. It is widely used as a pulpwood tree, and its heavy, strong, straight-grained wood makes it desirable for building material.

Balsam fir, the beautiful dark-green conifer of the Northern Forest, is found from Labrador west to Alberta, and its range extends down through New England and New York. It is not particularly large, averaging between 25 and 60 feet in height, and for years it has been prized as a Christmas tree because it is so symmetrical, fragrant, and retentive of its needles. Its aromatic needles are used for stuffing balsam pillows. Of no particular importance as a timber tree, it is much in demand for paper pulp because of the softness and light color of the wood.

Ponderosa pine was named by the botanist David Douglas, who first saw it in eastern Washington and suggested the term "ponderosa" because of its great size. The ponderosa pine frequently lives for three or four hundred years and reaches heights over 200 feet, with 5- to 8-foot diameters. Stands of this tree are scattered along the Pacific Coast and throughout the Rocky Mountains, where it is able to grow at elevations up to about 12,000 feet. One of our most important timber trees, it is widely used in construction work, and the better grades of wood are made into paneling and trim.

White spruce is one of the most widely distributed trees in the spruce family, growing throughout the Northern Forest from Labrador to Alaska. Its strong, resilient wood is utilized for lumber, and its resonant qualities make it highly desirable for sounding boards in musical instruments. Along with other spruces, it is a prime source of pulpwood for paper, and it is also used as lumber for general construction work. The white spruce usually grows to a height of 40 to 70 feet and reaches its greatest size in British Columbia and Alberta.

The colors on the map and the matching dots next to each tree show the principal areas where it usually grows.

Eastern white pine

Douglas fir

Longleaf pine

Western hemlock

Balsam fir

Ponderosa pine

White spruce

For nearly three centuries the attitude that dominated Americans' thinking about their forests was that the faster they could be cut down the better. It was a philosophy in which there was no tomorrow. To those first generations of immigrants the trees seemed to extend across the continent like an unbroken, eternally bountiful sea, so despite the fact that such prodigal forests did not exist in Europe, where wood of any kind was a luxury, the trees began to fall in uncounted numbers to the settlers' axes. The first sawmill opened for business in Maine early in the seventeenth century, and the towering white pines of New England were the first to go.

Trees so large that twenty yoke of oxen were required to haul one of them from the woods to the Maine wharves were loaded onto special ships that carried masts for the British Navy; these bargelike vessels had long, unbroken floors that could hold one-hundred-foot lengths, and the masts were slid aboard through a door in the stern, sometimes fifty masts to a ship. Houses, barns, forts, barrels, wagons—all were built of wood, and all consumed increasing quantities of it. During and after the Revolution, when the new nation depended so heavily on its foreign commerce, huge amounts of pine and white oak went into the growing merchant and naval fleets.

By the eighteenth century there were already a few signs that such profligate destruction of the forests had its inevitable consequences. Near the Cape Cod town of Truro the sand was beginning to drift ominously as a result of heavy timber-cutting and overgrazing, and the Great and General Court of Massachusetts was forced to consider legislation to regulate these practices. That colony and North Carolina also passed fire laws to protect their forests during the eighteenth century. There had, to be sure, been a few other voices crying in and for the wilderness: in his land-purchase contracts of 1681 the farsighted William Penn had required that one acre of trees be left for every five acres cleared in Pennsylvania, and the town of Newington, New Hampshire, had established a community forest in 1710. But these were the exceptions—the rule was wholesale abuse of the woodlands.

Within a generation of the Revolution the first sawmill began operating in New Orleans, and logs were being rafted down-river; a third of a century later the first wood-burning locomotive made its appearance, and by the Civil War thirty thousand miles of track had already been laid on millions of wooden ties in a feverish effort to connect every new settlement with the world back east. In 1830 the initial sawmill opened near Pontiac, Michigan, indicating that the logging business had leaped that far westward, and during the period between 1820 and 1870 the population of the United States quadrupled, multiplying significantly the number of people who needed articles made from wood. Timber demands continued to increase as the nation pushed ever toward the west; canalboats and farm buildings, new homes and cities, Conestoga wagons and plank sidewalks and furniture—everything called for lumber. And as the virgin timberlands of the Central Forest were decimated, the loggers turned south, chewing up the great pinewoods like locusts cutting a swath through a field of wheat. New towns were a-building everywhere, the sod hutters on the plains had to import every stick of wood they used, and the lumberjacks moved through Michigan and Wisconsin into Minnesota. As Stewart Holbrook wrote: "What they did not cut they and the settlers managed to set afire, creating

some of the most horrible disasters imaginable." Worst of all was the Peshtigo, Wisconsin, conflagration in 1871, in which fifteen hundred persons died and one and a third million acres of timberland burned. On into Montana, Idaho, and the Douglas fir and ponderosa country the axes and saws and peaveys went, until finally—although few people were aware of it at the time —the loggers reached the last outpost of virgin timber, between the mountains and the western sea.

This is not to say that the men who worked in the woods, and those who employed them, were wholly to blame; the entire nation had its eyes set only upon that elusive goal called Progress, and the woods seemed heaven-sent for exploitation, a bountiful provider of a multitude of things needed by the most restless, questing society ever known. At last, however, there began to dawn on the public consciousness the idea that the dwindling forests should be protected. One of the first voices to be heard was that of a Vermont lawyer and scholar, George Perkins Marsh, who once described himself as "forest-born." In 1864 he published a book whose significance has only recently come to be realized. He called it *Man and Nature: Physical Geography as Modified by Human Action*, and in it he created the concept of modern ecology, which sees the entire earth as a single entity that is affected by the complex interaction of cause and effect. Marsh came to his thesis in large part because of what he witnessed during the course of his lifetime. A year before his birth a forest fire burned off a mountain on the outskirts of his home town, and while he was growing up, he observed the effects of that disaster, realizing that it might take hundreds of years before there was again "a stratum of soil thick enough to support a full-grown forest." Later, as a diplomat, he saw the ruins of North Africa and other Roman provinces in Asia Minor and southern Europe, and he perceived how the environment had depreciated, exhausting the soil of its fertility. "Vast forests have disappeared from mountain spurs and ridges," he wrote, and he calculated that the annual silt from the denuded Apennines and Alps was enough to cover 360 square miles, seven inches deep. Denouncing the Roman Empire that had left a "dying curse to all her wide dominion," Marsh warned his countrymen that men were wrecking the earth and must one day suffer the consequences of their actions.

Nor was Marsh alone in his concern. Not long after the Peshtigo fire a group petitioned Congress to give attention to the "cultivation of timber and preservation of forests and to recommend proper legislation for securing these objects." In 1875 the American Forestry Association was organized for the promotion of forestry and timber culture; public interest in the subject increased; in the sixties and seventies homesteaders began planting trees on the plains to shade their homesites and protect their lands from the constant winds; and in 1876 Congress authorized appointment of a special agent in the Department of Agriculture who would be responsible for determining the facts about the need for timber and the means for preserving the forests, at the same time investigating what other countries had done "for the preservation and restoration or planting of forests." A beginning step had been taken, and within five years a Division of Forestry was established in the department. Agricultural colleges began offering courses in forestry, and in 1898 the New York State College of Forestry opened at Cornell University, followed two years later by Yale's graduate School of Forestry.

At the insistent urging of men like John Muir and Carl Schurz, some progress was made in setting aside large areas of government-owned timberland, but when Theodore Roosevelt became President and appointed a committee to study the nation's resources, the report he received was a gloomy one indeed. The wanton removal of trees had seriously reduced the water-conserving capacity of the soil; all over the country stream banks were eroding and washing away. And land that was not damaged by floods was being ravaged by flames, for the slash left on the ground by loggers after a big timber cut was an invitation to fire.

Roosevelt and the man he appointed Chief Forester, Gifford Pinchot, formulated a plan for the constructive use, preservation, and administration of the nation's forests, and in 1901 the President announced the beginning of what was to be a new era in American forestry. "The forest and water problems are perhaps the most vital internal problems of the United States," he said. "The fundamental idea of forestry is in the perpetuation of the forests by use. Forest protection is not an end in itself; it is a means to increase and sustain the resources of the country and the industries which depend on them. The preservation of our forests is an imperative necessity." An age of enlightenment had dawned for America's woodlands, and although there were to be many bitter fights ahead over public regulation of private lands, the guidelines set by Theodore Roosevelt and Gifford Pinchot eventually reversed a trend that had been systematically destroying America's trees since the days of the first settlements. In what may have been the most lasting of his contributions to the nation, Roosevelt set aside 132,000,000 acres of public lands, converting them into forest preserves.

Heightened interest in America's natural resources led, in 1916, to establishment of the National Park Service, and that same year a Migratory Bird Treaty reflected the public's growing concern over dwindling wildlife species. Other positive legislation was enacted, but one of the greatest boosts for forestry came about as a result of economic catastrophe: in 1933 the Civilian Conservation Corps was created by Congress to provide jobs for thousands of young men who were unable to find work during the depression. At the high point of this program, which was intended to build up the national resources, more than half a million men were enrolled in CCC camps in activities that substantially improved state and federal woodlands at the same time they gave training and experience in forestry to many Americans. Another important program resulted from the creation in 1933 of a Soil Conservation Service, which encouraged the reforestation of lands for soil and water conservation.

Implicit in all the prior history of forest depletion was the warning sounded by George Marsh in 1864. Seventy years later some of his gloomiest predictions came to pass, when drought and windstorms reduced millions of acres of plains land to dust, and brown clouds of topsoil dimmed the sky as far east as Albany, New York. Great portions of the land had turned to hardpan or sandy dust, for where there had been trees, there were now millions of stumps, and the once rich topsoil blew away or ran off with the spring floods. The toll of destruction wrought by misguided farming and logging practices was reflected in the distress of families no longer able to scrounge a living from worn-out, eroded land that had been unsuitable for agriculture in the first place. In human terms the effects of the dust bowl were incalculable; it was a time when thousands of Americans were refugees in their own land, homeless and jobless through no fault of their own, victims of the lack of vision and understanding of previous generations. Nor has most of the land recovered: millions of acres that will not support vegetation remain unusable because of the centuries of waste and neglect.

Apart from federal and state lands that were coming under some kind of control during this period, there remained the continuing problem of small private timberlands, which even in 1970 constitute more than half the total forest acreage in the United States. Beginning in the 1950's increased attention was given to means of improving and properly managing these lands. The Tree Farm program, under which privately owned timber stands were committed to the regular production of forest crops under sound management practices, had expanded by 1970 to nearly thirty-five thousand tree farms comprising some 74 million acres. In 1969 800 million trees were planted under various reforestation programs; more than 90 per cent of all forest lands were under organized fire protection; and, as we shall see in the following pages, numerous paper and lumber companies were adopting imaginative techniques for more effective management and utilization of timberlands.

Forest acreage, which had reached an all-time low immediately after World War I, began, by the late 1930's, to increase. Between that time and the present, conservation practices and the new ideas of forest management, enhanced by scientific breakthroughs in many related fields, have achieved a healthy balance between what must be harvested and what must be preserved for the future.

From the experience of the past the modern forest products industry has learned many lessons—among them that of proper harvesting. Instead of the old practice of cut-and-get-out, today's forest manager is compelled to think in terms of the future as well as the present, of the continuing need for more and better trees, and of the necessity for intelligent supervision and regeneration of woodlands. The principles involved in the care and improvement of trees are known as silviculture—a practice that requires knowledge of the life history and characteristics of trees with an understanding of environmental factors; its goal is to reproduce and manage forests in order to obtain a continuous output of crops. America's primeval forests were created with no help from man, but what has been learned about various aspects of forest growth now makes it possible to obtain better yields of desirable timber in a shorter period of time than is possible in a natural, untended forest. Implicit in such an approach is the wise use of man's skills to assist nature.

The objective of forestry today is sustained yields as opposed to the opportunistic harvesting of a one-time crop, and the techniques employed fall into several categories, whose application depends upon whether the woodland is a mature stand or a young growing stand. After careful studies have been made to determine in advance what procedures are to be followed in a forest, one of the first steps undertaken is timber-stand improvement, a process designed to improve the over-all quality of the stand by removing poor and unsound trees in order to improve the growth rate of more desirable plants. An overcrowded forest is one with an unsatisfactory growth rate because the roots as well as the branches of the trees are in competition with those of their neighbors, restricting their development. If the forest has too many old, heavily branched trees, known as wolf trees, they suppress young saplings at the time they would normally be growing most rapidly. Frequently, too, undesirable species are present, crowding out more valuable trees. One phase of timber-stand improvement is the cutting of these trees in order to liberate those the forest manager regards as more promising for the future. In a vigorous young stand periodic thinning is required to prevent overcrowding; for example, on sites where the land has been heavily seeded, as many as 20,000 little seedlings may be found to the acre, but after twenty-five or thirty years only 400 to 800 trees should remain on the site. And by the time the trees are full-grown, only 150 or 200 healthy, evenly spaced trees may be left.

The drawing on page 89 is a dramatic example of what good management can achieve in a commercial forest. Both sections of western white pine are shown about one-fourth their typical size, and if we could count the tree rings in the drawing at the top of the page, we would see that that tree was 280 years old when it was cut. Its life began in 1690, and from the evidence of the wide rings near the center, we judge that it had a good climate for growth as a seedling. Then the tree's growth slowed down markedly. The next rings are tight and

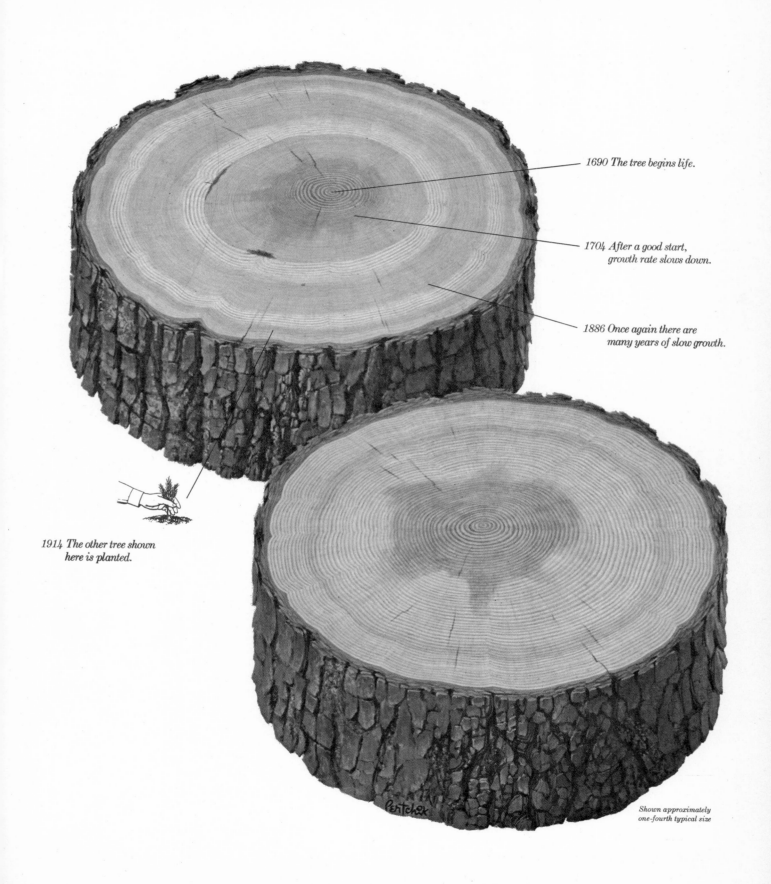

1690 The tree begins life.

1704 After a good start,
growth rate slows down.

1886 Once again there are
many years of slow growth.

1914 The other tree shown
here is planted.

Shown approximately
one-fourth typical size

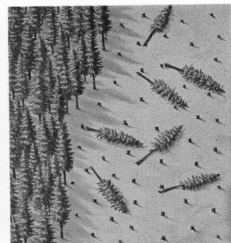

Clear Cutting

Trees like the Douglas fir need full sunlight in order to attain proper growth.

The accepted way of harvesting Douglas fir is to cut all marketable trees in a block of 100 acres or less, leaving standing timber between the logged areas (top, right).

In time, seeds blow in from the surrounding forest, and seedlings take root, as shown at bottom, left. The new stands are thinned periodically to promote growth.

About forty years after the cut, the new stand is mature enough to provide seeds for a neighboring block, which is then cut. A few decades later the first block is harvested (illustration below at right).

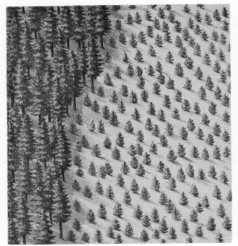

small, indicating that the trees near this one were growing at about the same rate; their branches had become entangled in the competition for sun, and their roots had intertwined in the search for water and nutrients. For years the development of this tree was retarded, and not until some natural force—a fire, disease, or weather—intervened to eliminate some of the competitors did the tree commence its normal growth again. Then there was another long period of slow development when neighboring trees crowded in on it again; then an increase; and so on. The point is that it required nearly three centuries for this particular tree to reach the same diameter as the one at the bottom of the page. By contrast, the latter took only fifty-six years to attain

the identical growth, and the reason was proper forest management. By following the methods of nature, it is possible to grow trees faster in a shorter period of time. Continuous thinning and improvement of the stand give each tree adequate room to grow, and the growth rate remains virtually constant throughout the life of the tree.

Today, four principal methods of harvesting trees are in common use. One is called selective cutting—a technique often employed in a forest populated with trees of all ages, whose purpose is to remove large, old trees as they reach maturity, thus improving the quality of the rest of the stand. Left to its own devices, nature does much the same thing, but the process is considerably

Seed-tree Cutting

Like the Douglas fir, southern pines need full sunlight for vigorous growth, and their seeds are also carried by the wind.

In seed cutting, all trees are removed except for four or five per acre (top, right). These act as seed trees, and are allowed to remain until seedlings are established.

When seedlings have grown enough so that they are reasonably safe from fire (five or ten years), seed trees can be removed in their turn (illustration at lower left).

After about twenty years the new stand is thinned to prevent overcrowding. Under favorable conditions, the pines can be harvested about thirty years after original cut.

slower, and the results not always satisfactory. By choosing trees carefully for cutting, with an eye to the way they may affect the future growth-potential of nearby trees, selective cutting lets sunlight into the forest gradually, and for this reason it is especially suited to use in hardwood stands of shade-tolerant species. It reduces the possibility of windfall and fire, makes reproduction relatively certain, and assures continuous protection and food for wildlife.

Another method of harvesting is known as the shelterwood system, by which trees are removed at several stages of development, with the dominant and inferior ones being cut first. Some species require a certain amount of shade and shelter during their early years,

but are also slow to grow if this protective cover remains. In the shelterwood process the first, or preparatory, cutting is designed to eliminate the dominant and defective trees, letting in enough sunlight to encourage new growth. The next step is called a seed cutting, which may remove half or more of the remaining trees immediately after a year that has produced a particularly good crop of seeds. What remains after this stage are the strongest and best mature trees, which continue to provide some protection for the new seedlings until they have attained sufficient size so that the older trees can be cut in the final removal stage. Especially effective with red, white, ponderosa, and southern pines and in some oak forests, the shelterwood system has the ad-

vantage of reducing crowding and preventing brush from overtopping the seedlings. It also permits natural reforestation from seeds of the best trees, affords these seedlings protection as long as they need it, and—as in the selective-cutting method—makes for a gradual change in forest conditions, enabling young trees to adjust to a slowly altered environment.

An entirely different method of harvesting is the one known as clear-cutting, which is most often practiced in timber stands where all the trees are approximately the same age. This technique is used with species such as Douglas fir, jack and lodgepole pine, western larch, and black spruce that reproduce easily in direct sunlight. Immediately after the area has been clear-cut it is seeded—sometimes naturally, from adjacent stands that are left for this purpose and for protection; sometimes by hand; or sometimes, in large areas, by seeds broadcast from an airplane. One advantage of the clear-cutting method is that it produces even stands of trees that have no competition from older growth. To reduce the hazardous effects of leaving extensive acreage subject to erosion or drying by sun and wind, clear-cutting is usually carried out in confined strips or blocks, and uncut timber is left on the windward side of the harvested area.

The fourth method by which trees are frequently taken out of a forest is called the seed-tree system, in which something like 10 per cent of the best seed-bearing trees are left on the site to provide the seeds necessary for the next crop of trees. In general, about five of these mature seed-trees are allowed to remain on each acre of land for a period of five or ten years, or until a new generation has been established, at which time the old trees are cut. This system may be practiced successfully with several of the fast-growing southern species, like longleaf and loblolly pines.

In fact, the last three methods described are widely used in the pulpwood industry to cut fast-growing trees that are often ready for harvesting thirty years after they have been seeded. Selective cutting has been the more normal practice in hardwood stands, where trees take much longer to mature. Here, the principle followed has been to take out a relatively few trees, creating small openings in the stand to release young growth, and then to make subsequent selective cuttings at regular intervals of five or ten years. The over-all effect of this method is to encourage the climax trees, of course, at the expense of the pioneer species. In other words, if the paper birch is gradually removed from a New England hardwood forest, there is very little regeneration of that species; instead, the shade-tolerant trees are encouraged, and beech and sugar maple eventually dominate the woods, to the virtual exclusion of the paper birch. When it is desirable to reproduce those trees that are intolerant of shade—birch, yellow poplar, or black cherry, for example—it is argued that the only practicable method to follow is clear-cutting, since these species cannot flourish in the shade of the climax trees. Another argument that has been advanced for strip or block clear-cutting in hardwood forests is that water that would normally be utilized by mature trees in these areas is thereby allowed to move into drainage systems and reservoirs. A study made in one eastern hardwood forest indicated that the water yield during the first year after a clear-cutting of one hundred acres would be nearly 14 million gallons.

Which brings us to what the objectives of the forester are. There is no question but what properly managed forests produce a greater volume of better timber than do those in which there has been no planning for the future. The concept of renewing forests came into being some seven hundred years ago, when the feudal masters of Europe began to realize that their forests and game habitats were vanishing and sought to perpetuate their resources by restricting the removal of trees. Today, forestry is more nearly a science, whose goal is the management of forest lands to satisfy the needs of man, and both government and industry have come to the realization that forests must serve man in ways other than fulfilling his requirements for wood. Some forests, it is clear, must be farmed for trees; others must be preserved as wilderness and recreational areas; and still others must be utilized for both purposes, for it has been predicted that by the year 2000 the United States will need 300 million more acres of timberland than it has today just to supply the vast array of forest products it consumes. At the same time the pressure for recreational land will have increased threefold to sixfold.

Yet merely adding new forest acreage—if that could be done—cannot in itself answer the problem of the ever-growing needs for timber. The encouraging experience of the past few decades suggests that partial solutions are already available through improved techniques of managing forests, the breeding of superior trees, and making better use of the timber that is cut. The regular improvement of existing stands is already extensively practiced on government tracts and on lands owned by large industrial concerns, but it is essential that similar techniques be applied to forest lands in private hands, which constitute half of the total woodland in the United States.

On page 98 we will examine some of the ways in which more efficient utilization of timber has been accomplished. What is perhaps more heartening is the degree of success already achieved in developing new strains of trees that produce more usable wood in less time. Any form of artificial regeneration involves what are essentially genetic choices, and the most important decision to be made in reforestation is which species will be planted in a new stand. It must be suited to the particular soil, climate, and environment that prevail on the site, and ideally the species should represent the best characteristics to be found in hardy local strains. Specialists in tree genetics begin with the assumption that outstanding trees in a particular environment have desirable characteristics for future reforestation in that area. If it is found that they are capable of transmitting those characteristics to their progeny under carefully

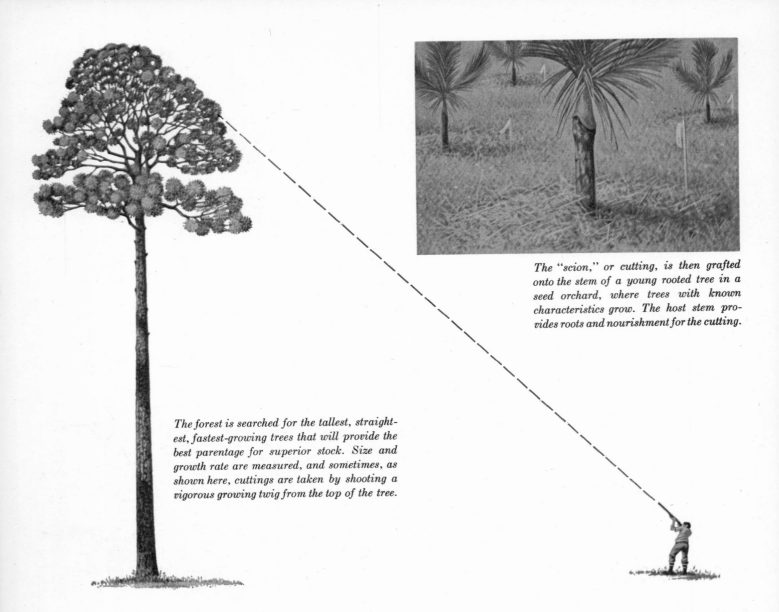

The forest is searched for the tallest, straightest, fastest-growing trees that will provide the best parentage for superior stock. Size and growth rate are measured, and sometimes, as shown here, cuttings are taken by shooting a vigorous growing twig from the top of the tree.

The "scion," or cutting, is then grafted onto the stem of a young rooted tree in a seed orchard, where trees with known characteristics grow. The host stem provides roots and nourishment for the cutting.

controlled conditions of reproduction, their seeds can be gathered to supply nursery stock, and in time the most vigorous specimens can be selected for continuous improvement of the species. As may be imagined, the collection of good seeds from forest trees is no easy matter. The most satisfactory producers of seeds are usually the dominant trees in a stand, but their seeds are not easily accessible. In general, seeds to be used for reforestation should be collected as near as possible to the planting site, and the best seeds are obtained in unusually productive years, just after the cones have matured, but before they have begun to shed their seeds. Because the collection of large quantities of seeds from a living forest tends to be inefficient, it is much more sensible to establish orchards consisting of the best strains and manage them intensively for seed production.

Apart from this carefully selective nursery procedure, there is the possibility of actually breeding improved strains of trees—a long, painstaking process involving the careful choice of certain desirable characteristics that are to be bred into future plants. Elaborate tests are made to determine the superior types for a particular purpose, and these are crossed with others to yield offspring that have a high representation of the desirable characteristics—trees that are faster growing, taller, larger, and more resistant to disease. The new hybrid plants are set out in orchards and become a source of seeds for the production of planting stock. As genetic improvement has become more sophisticated, cross-pollination has produced new strains of trees that could not have developed naturally—making it possible, for instance, to combine trees that grow in widely separated geographic locations. It has been found that some of these new hybrids grow considerably faster than either of their parents, and they may also reveal entirely new and unsuspected characteristics.

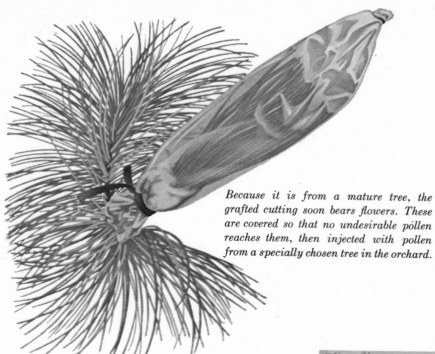

Because it is from a mature tree, the grafted cutting soon bears flowers. These are covered so that no undesirable pollen reaches them, then injected with pollen from a specially chosen tree in the orchard.

After the seeds form, the cones are picked, and the seeds are extracted by mechanical shakers, to be kept in cool, dry storerooms.

Seeds are planted in nursery beds, to grow under carefully controlled conditions. After a year they have sturdy roots and well-developed foliage and are transplanted.

Superior seedlings from the orchard replace a stand of mature trees that have been harvested. Planted in rows with enough space for ample growing room, the new crop grows faster, is more resistant to disease and insects, and yields more wood of better quality.

In a commercial forest it is important to remove each tree while it is still healthy so that it can be converted into useful products before it deteriorates. Unless it is properly thinned, a thick stand of young, even-aged trees quickly stagnates as the growth rate slows down, because the competition for sunlight, water, and nutrients is so fierce. Without careful management this competition gradually results in the dominance of stronger trees and the suppression or death of weaker ones. To prevent this, damaged or repressed trees are removed as early as possible; the forester anticipates their natural loss and has them cut while they can still be sold for pulpwood, posts, or other uses. It is important that the crowns of trees in a commercial stand be close enough to encourage natural pruning by shading the lower limbs so that they drop off, and the density of trees should be maintained at a proper level so that they can make optimum use of light, moisture, and mineral nutrients. As these illustrations indicate, thinnings made at various stages of a forest's development are carefully calculated to make efficient use of each successive cutting at the same time that growing conditions are improved for the remaining trees.

Within its first eight to ten years a Douglas fir may be harvested for a Christmas tree. Thousands may be cut to "thin" a forest.

At thirty to forty years trees may be cut for pulpwood. Deprived of light, the lower limbs of this tree have now begun to drop off.

At age fifty or sixty a tre may be 100 feet in heigh Now it can be cut for pu or for a telephone pol

100-year-old tree may ave some "clear" wood nmarked by knots—good or plywood or lumber.

After a hundred years trees grow at a slower rate. In virgin forests some Douglas firs survive for hundreds of years and reach heights of 300 feet.

No matter what its age, a declining tree is easy to spot. Its branches are dying, its bark scaling, and rot has begun to spread through the trunk.

Upper branches fall away, leaving a bare spike. At last the tree is weak enough to be blown down in a high wind, or it may topple of its own weight.

J.KUNZ

At the turn of this century as little as one-third of a tree might be converted into a handful of useful products; at the sawmill square lumber was cut out of round trees, and the waste was burned or left to rot. In those days, lumber accounted for 90 per cent of the use of logs. Today, about 50 per cent of all timber cut goes into saw logs for building, but each American consumes an average of 575 pounds of paper every year, and enormous quantities of wood go into veneer and plywood, fuel, poles, posts, and some five thousand other products. The demand for lumber and paper seems insatiable, and it increases by 4 per cent every year.

Modern improvements in processing machinery make possible a much more efficient utilization of each log: improved saws minimize waste in cutting; barkers strip the bark from logs so that there is virtually no loss of valuable wood fiber, and the bark itself is converted into valuable products; machines known as chippers take pieces of wood once regarded as useless and reduce them to material that is made into pulp for paper and cardboard. This idealized drawing shows how a log is cut for lumber and other purposes; as it is sawed, the slabs and many other pieces, including edgings and trim ends, go to the chipper.

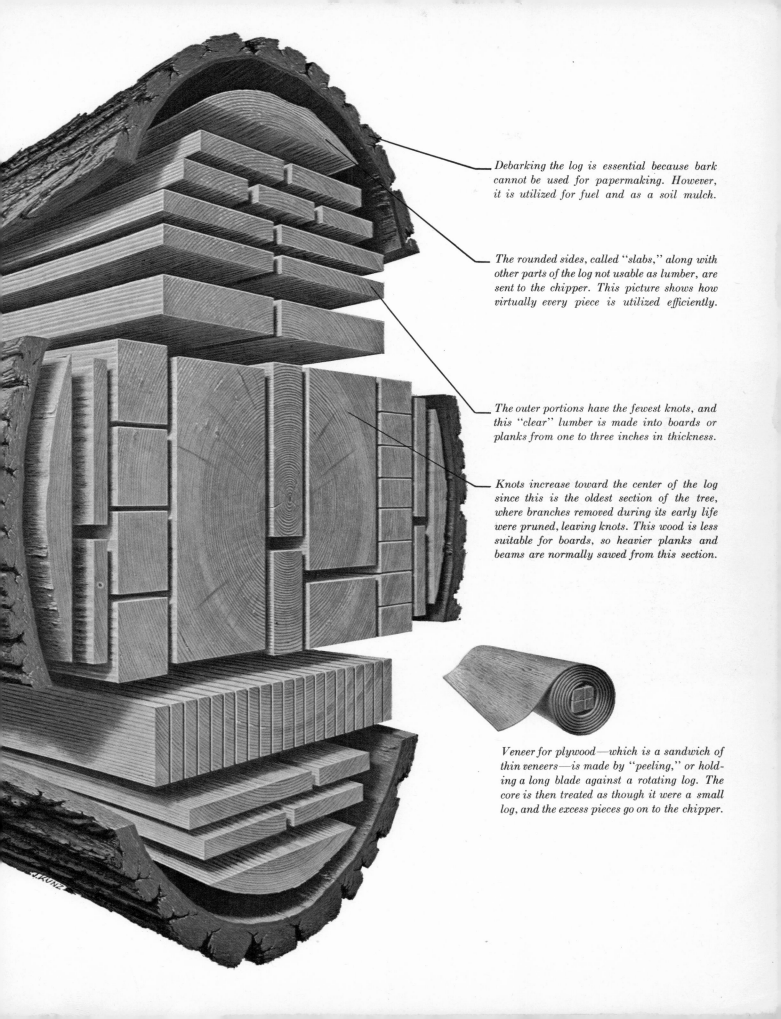

Debarking the log is essential because bark cannot be used for papermaking. However, it is utilized for fuel and as a soil mulch.

The rounded sides, called "slabs," along with other parts of the log not usable as lumber, are sent to the chipper. This picture shows how virtually every piece is utilized efficiently.

The outer portions have the fewest knots, and this "clear" lumber is made into boards or planks from one to three inches in thickness.

Knots increase toward the center of the log since this is the oldest section of the tree, where branches removed during its early life were pruned, leaving knots. This wood is less suitable for boards, so heavier planks and beams are normally sawed from this section.

Veneer for plywood—which is a sandwich of thin veneers—is made by "peeling," or holding a long blade against a rotating log. The core is then treated as though it were a small log, and the excess pieces go on to the chipper.

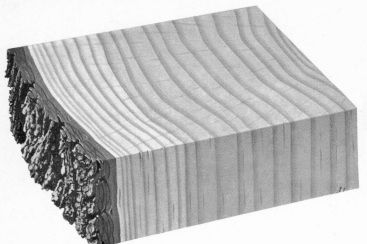

White pine is a softwood used in home construction and for many other purposes, including masts, matches, boxes, and crates. It cuts easily, polishes well, and warps little.

White oak makes good barrels because it is resilient, durable, and impermeable to liquids. It is nearly twice as dense as pine and is used for flooring and cabinetwork.

Hard maple was used by the Romans for spears and lances. Uniform texture and hardness make it ideal for flooring and bowling pins, for turning on lathes, and for spools and bobbins.

Baldcypress is weather-resistant even without treatment and was widely used for ties for early railroads. It is desirable wherever wood will be in prolonged contact with water.

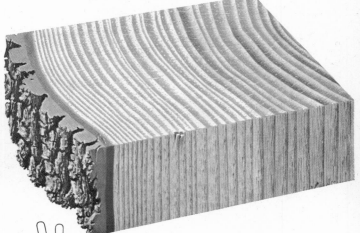

Black walnut, because of the beauty of the heartwood grain and its good machining properties, is a choice hardwood for furniture and interior paneling. It is harder than oak.

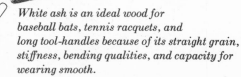

White ash is an ideal wood for baseball bats, tennis racquets, and long tool-handles because of its straight grain, stiffness, bending qualities, and capacity for wearing smooth.

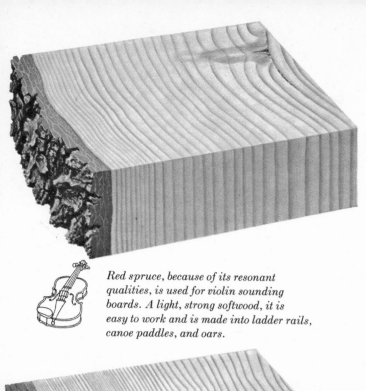

Red spruce, because of its resonant qualities, is used for violin sounding boards. A light, strong softwood, it is easy to work and is made into ladder rails, canoe paddles, and oars.

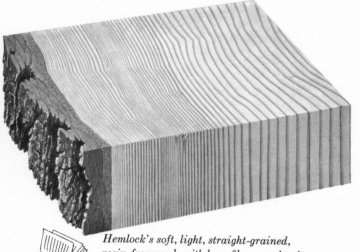

Hemlock's soft, light, straight-grained, resin-free wood, with long fibers, makes it an important species for paper pulp, as well as for structural lumber, plywood, boxes, and barrels.

Hickory is unsurpassed as a hardwood for making handles of impact tools like axes and hammers. Its hardness, strength, toughness, and resiliency also make it suitable for skis.

The astonishing range of products man has contrived to make from trees is impossible to catalogue in a short space, for more than five thousand objects in common use originate in the forests of North America alone. Without considering the array of wooden items we see around us on every side, they range from wild plums that are made into jam for the breakfast table to shatterproof glass that is a by-product of pulpwood. From the sap and resins come maple syrup, adhesives, varnish and printing ink, shoe polish and explosives. From bark we derive acids and dyes, oils and drug products; from leaves we get our Christmas decorations and various oils. Pulpwood is converted into all kinds of paper, board, rayon, plastics, imitation leather, alcohol, and sausage casings. Sawdust from the mill becomes bedding for dairy cows, is made into artificial wood, linoleum filler, and briquettes for outdoor cookery. From the stumps of trees we make charcoal, pitch, and tar products, and from the roots come oils, tea, and brier pipes for the smoker. Today, greater utilization of every part of the tree is an important aspect of forest conservation, with the result that more of the residue is consumed, and inferior species and poorer grades of trees can be used.

Among the many considerations that enter into the choice of wood for a specific purpose are weight, density, the moisture content, stiffness, toughness, and the presence of knots and resin. A wood is judged by its tendency to shrink or swell, by its strength, by its reaction to paints and stains. No two woods have an identical structure. The fibers of a softwood tree and the vessels of a hardwood tree conduct water when the tree is alive; eventually these slender cellular units grow together to form the substance we call wood, and the way they do so imparts certain characteristics to it. As a result, white pine and yellow poplar are easily worked; hickory and ash have great bending strength; sugar maple, white oak, and persimmon are extremely hard; the cedars and redwoods contain oils and other substances that resist decay organisms. A resinous wood does not paint well, a dense wood holds nails better, and so on. The so-called hardwoods are generally favored for making furniture and tools, while softwoods are used most frequently for construction materials and for paper products.

In any given year several million people visit one or more of the 154 national forests in the United States, not to mention the hundreds of thousands who spend some time in the private forests of the nation. All of them —hikers, hunters, fishermen, campers, canoeists—are drawn toward the woods for some special reason, yet it is doubtful if more than a handful of these millions of people seeking recreation or relaxation or solace see the forest whole, as the "web of life" it truly is. The casual hiker or camper may think of it as a cool, moist sanctuary, a retreat where he is occasionally aware of the quick flash of wings as a bird flies across his path of vision, or where he may catch a glimpse of a wild animal. The bird watcher is on the lookout for particular species, the botanist for unusual ferns or shrubs or trees. But the forest is far more than any of these things: it is a vastly complex community of plants and animals, all mutually interdependent, each performing some vital function that makes possible the existence or survival of others. Now and then we are obliged to consider the realities of this interdependence: what, for example, becomes of the alligators and other reptiles in the Florida Everglades if their water supply is depleted or if the intrusion of man becomes too much for them? What happens to a forest if the population of chickadees, nuthatches, or woodpeckers is suddenly reduced so that the insect population gets out of control? What are the effects on the surrounding countryside when more than 100,000 acres of ponderosa-pine forest suffer smog damage, as they have in the mountains near Los Angeles?

The balance of nature is a fragile line, indeed, and has always been so—even before the influence of man became such a disturbing factor. Any time it is upset, certain members of the natural community are affected, and the result can mean the extinction of certain plant or animal species. We may never know what caused the demise of the mammoth or the saber-toothed tiger in prehistoric times; their extermination may have come about because of great climatic change or competition from other species or an invasion of some new disease. But in more recent eras man has too often been the agent of destruction, either as a predator or as the unwitting agent of change which brought about certain conditions that proved intolerable to a species.

A classic example that is often cited to illustrate what can happen when the balance of nature is altered took place in the Kaibab Plateau of Arizona after 1907, when the government launched a campaign to eliminate the coyotes, pumas, wolves, and other predators from the region in order to make it into a game preserve for deer. This 700,000-acre forest of Engelmann's spruce, ponderosa pine, and Douglas fir was, in the words of one nineteenth-century visitor, "the most enchanting region it has ever been our privilege to visit," and for centuries it had been a hunting ground for Indians who killed the mule deer for food and skins. When the government stepped in and prohibited deer hunting, it was estimated that there were about four thousand deer in the forest; as part of the program, their natural enemies were systematically removed, and over a period of twenty-five years thousands of predators were killed by government hunters. By 1924 success seemed to have been achieved: there was a fantastic increase in the size of the deer herd, until the numbers approached 100,000. But this, as it turned out, was vastly beyond anything the region could support. The delicate balance that had been maintained by the predators had been completely upset; the deer population multiplied to such an extent that the region could not possibly support it, and the consequences were tragic. Suddenly the deer were threatened with wholesale starvation, and during the winters of 1925 and 1926 more than half of them died for lack of food. Those that were left devoured every available twig and leaf until, in the words of a contemporary account, "the whole country looked as though a swarm of locusts had swept through it, leaving the range (except for the taller shrubs and trees) torn, gray, stripped, and dying." Some thirty-five years after the program had been launched to "help" the deer herd, the population was down to about eight thousand, most of them sickly and undernourished, and the composition of the forest itself had been radically altered. Certain forms of plant life had disappeared altogether, and others had suffered severe damage; the deer had browsed

A sharp rat-a-tat-a-tat in the woods indicates that a woodpecker is at work, drilling into the bark to get at the insect larvae that have burrowed beneath it. In the woods birds are most often to be seen in or near a sunlit clearing, especially near berry bushes; the migrating birds frequently stop over to feed in such locations.

*Red-banded leaf hoppers suck
the sap from leaves in the
canopy, an area where thousands
of other leaf-eating insects feed.*

the young trees so heavily that few vigorous specimens
matured to replace the old growth, and gradually the
once heavily forested terrain changed to a more open
area in which grasses prevailed.

The lesson of the Kaibab forest is that man is and
must remain a part of nature, and what change he
brings about must be in harmony with the basic laws of
nature. A forest abounds with life, the basis of which
is plants—the food producers—that take raw materials
provided by sun, water, earth, and soil, and convert
them into living matter. Yet without the activity of
animals and other organisms, plants would long ago
have stored away in the earth the planet's available resources of carbon and then perished. It takes animal
life to consume the plants and release the carbon in
order to keep the cycle of life in constant motion. There
is, in America's forests, a wealth of wildlife, and the
character of the fauna in any given area depends, as
that of the flora does, upon the amount of available
light and water, the climate, and the nature of the soil.
Every region has its quota of invertebrates and verte-
brates, its distinctive varieties of cold- and warm-
blooded creatures, living in what are called ecosystems.
The term refers to a particular physical environment
and the community it supports—a situation that might
be likened to an aquarium or fish bowl. There, water
constitutes the physical environment. It contains a
variety of dissolved minerals; light enters the tank,

*The shrew, a tireless floor-
burrower, feeds on insects, worms,
and rodents beneath the
protective cover of shrubs.*

The white-breasted nuthatch
is one of many trunk travelers
that swoop under the canopy to
prey on the insects living there.

Flying squirrels and many
songbirds are branch-dwellers,
nesting in the understory, out
of reach of the ground predators.

Each variety of woodpecker
(this is the downy) specializes in
a different part of the tree,
digging insects from the bark.

The rose-breasted grosbeak
is one of the songbirds that
nest in the shrub layer, protected by
the dense thickets there.

making it possible for green plants to grow; and fish feed off the plants. Just as the aquarium is a self-contained system that is capable of surviving for a time entirely on its own, so, on a much larger scale, is a lake, and so is a forest.

Four principal elements in an ecosystem make it possible for life to exist and to renew itself. The first, of course, is the physical environment itself, which receives light and energy from the sun. Then there are the green plants—organisms that transform the sun's energy and inorganic materials into sustenance for the third factor in the ecosystem—animal life. Herbivores of all kinds consume the plants, converting them into animal tissues and energy, and many of those herbivores are then consumed by carnivores. The fourth element in the ecosystem—and the one that makes it possible for the initial organic matter to be returned to the environment for re-use—is decomposition, the action of fungi and bacteria, which maintains the carbon cycle.

As an illustration of the interdependence that characterizes any natural community, consider what is necessary to support certain levels of animal life on one single acre of grassland. One study indicated that such a community had nearly six million grass plants, the primary food producers. These plants supported 700,000 herbivores—most of them insects—which fed off the grass and absorbed its energy. At the next higher level of this food chain were more than 350,000 carnivores—

spiders, ants, beetles, and other small creatures—feeding on the herbivores. And at the very top level were three larger carnivores—two moles and one bird—that fed on the lower forms of life and owed their existence to the six million grass plants at the base of the pyramid.

A forest community is infinitely more complex than an acre of grassland, and one way to think of it is as a series of horizontal bands or levels, arranged from the tops of the trees to the ground on which they stand. Each of these levels constitutes a habitat where different types of animals and plants live. The top level is called the canopy, which is formed by the crowns of the tallest trees. If the trees are fairly widely spaced, enough sunlight penetrates the canopy and reaches the layers below it to encourage an abundance of life there; if the trees are close together and the canopy dense, sunlight is shut out to a great extent, making it much more difficult for plant and animal life to prosper at the lower levels. Since light in the canopy is intense, this is where photosynthesis is carried on most actively; but the very intensity of the light, as well as the harsher weather conditions that prevail there, makes the canopy a rather inhospitable place for most forest animals, which are likely to seek shelter from the sun, wind, and rain at lower levels. Although certain birds—among them crows, hawks, and eagles—nest in the topmost branches of the forest, conditions are more acceptable and the food more plentiful beneath the upper branches, and this is the domain of thousands of leaf-eaters—aphids, leaf hoppers, caterpillars, beetles, and leaf miners. Their presence in large numbers attracts insect-eating birds, spiders, and other predator insects, and squirrels are also active in the canopy because seeds and nuts are plentiful there.

Below the canopy is the understory, a level of vegetation made up of smaller trees pushing upward toward the light. Some are prevented from doing so and eventually die; others succeed in the attempt and finally take their place as dominant trees in the forest. The understory has its own population of animals, birds, and insects, many of which prefer to nest there or find the feeding conditions more promising than at other levels. A study made in New York State indicated that different types of warblers occupy rather well-defined areas in the woods: one variety resides in the canopy, another in the understory, still others in the shrub layer or near the ground. This is their method of reducing competition and thus enabling a larger variety of birds to occupy a given region.

The next level is called the shrub layer. These woody plants are found less in a dense conifer forest than in a young deciduous woods where a good deal of light penetrates the canopy, so that the floor is often clogged with shrubs and vines of all kinds, which offer still more nesting and feeding opportunities for other birds, insects, and animals. This is the haven of many songbirds, which rely on the berries and seeds of shrubs for food, and of small animals such as chipmunks, shrews, and mice, which forage in the branches and burrow in the ground.

Under the shrubs is the herb layer, made up of green plants with soft, rather than woody, stems. Here, amid the low-lying wild flowers, grasses, ferns, mosses, and fungi, the little groundbirds flit in and out, and this is the habitat of mice, insects, snakes, toads, and larger predators that lie in wait for them.

At the bottom is the forest floor, where the great accumulation of autumn leaves or rotting needles piles up, mixed with twigs, branches, flower petals, and animal refuse, to be turned into humus by the activity of earthworms, millipedes, ants, and the fungi.

Life in these separate forest strata varies greatly from season to season, of course, but it is a mark of our increasingly urban society that few of the millions of city dwellers have the opportunity to realize or appreciate how profound the great cyclical changes are. At their extremes, they range from the arctic ferocity of winter to near-tropical heat in summer, and from the miraculous rebirth of spring to the flaming change of autumn, when the trees and other organisms prepare for winter once again. Each of these seasons represents a distinctive and separate phase in the life of plants and the myriad of creatures in the forest community, whose pattern of existence is dominated by nature's neverending process of change.

Near shady patches and decaying logs there are often specks of vivid red—tiny plants called "British soldiers." Among them are often "pixie cups." Both are species of lichens, a hardy plant form.

The anglewing is one of the few butterflies of early spring; most others are in the process of changing from crawling creatures that chew leaves to flying creatures that sip nectar.

The marsh marigold, bloodroot, and hepatica appear in early spring when ample moisture is present and sunlight reaches them because the trees' leaves are not fully formed.

Perhaps the most exciting time in the forest is spring, when so many plants and animals seem to come to life once again after the long winter of inactivity. Weather rules the forest, and as soon as the sun and warm winds of spring put a stop to the nightly freezing and start drying out the earth, wild flowers carpet the forest floor in a sudden burst of color before the trees develop their leafy canopy, a delicate green haze shows up on the tree branches, pale yellow and green sprouts appear everywhere, and the insects move in, as if on signal, to begin feeding on the pollen of young flowers. As long as the wet weather holds, the chorus of thousands of frogs and spring peepers fills the air. Newly hatched caterpillars start to munch on tender leaves, animals peer from their burrows to sniff the soft, warm air, and each week sees changes in the bird population as arrivals wing in from the south—to stay for the summer or to pause briefly before continuing on to a more northerly habitat. Birds and animals begin establishing their "territories," where they will raise and feed their young, and there is continuous rustling and foraging about as the many different kinds of birds begin the annual business of nest building. The strength of the forest is reckoned in water, and during the warm days and cool nights a rush of moisture from the ground carries water and nutrients upward to the buds until they burst open into leaves or flowers.

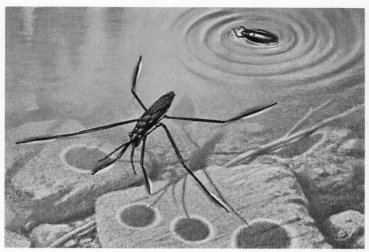

The forces that renew life are constantly at work in a dead log. The unseen decay bacteria and mosses and fungi help break up dead organic matter into its constituent parts.

Here the water strider skates on a pond with piles of oily hairs on its feet. The whirligig beetle has bi-focal eyes, half of which see under water, half over it.

Beneath the rocks the forest floor teems with life; above are shown pill bugs and sow bugs, which are relatives of the crawfish, and centipedes, which customarily rest by day and hunt by night.

In springtime the rabbits come out to munch the succulent young leaves. Their eyes, on the sides of their heads, enable them to see in both directions at once.

New leaves, formed within buds the previous fall, burst forth be-cause of increasing warmth and moisture and lengthening days.

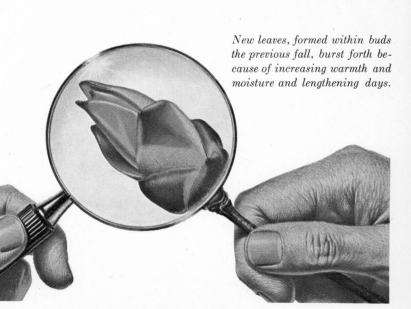

At the opposite end of the scale from spring's bustling activity is winter—the forest's quiet time, when plants and the animals that must remain there have to endure extended periods of cold and long food shortages. Most of the birds migrate southward, on journeys that take some of them thousands of miles distant, and a number of animals hibernate—some in a status that is only a thin borderline between life and death. Yet there is always activity in the woods in winter: the chipmunk emerges now and again from his sleep to collect more seeds; squirrels scamper about in search of nuts they have buried; the bear awakes occasionally from his torpor; and other animals, like the deer, forage throughout the winter for twigs and buds that are not covered by the snow, while the fox and weasel keep up their relentless hunt for food. Down beneath the snowdrifts life continues on the forest floor: there are insects in a state of suspended existence, in the form of eggs or cocoons; the fungi and bacteria stay alive under the insulating cover of snow and continue to decompose leaves and animal litter. Perennial wild flowers that will bloom in the spring grow their buds, drawing on food they accumulated the year before in roots or bulbs. The conifers and some small plants continue the process of photosynthesis on warm winter days, and the roots of most trees grow longer, in order to supply the need for minerals and water with the arrival of spring.

A bird equipped for winter is the ruffed grouse, which develops a comblike growth on the sides of its toes that helps it walk on snow. To elude foes, grouse dive into snowbanks.

Some animals are specially adapted to traveling in snow. These are the broad tracks of a snowshoe hare, whose feet are twice as wide as those of the jack rabbits that live in much warmer climates.

Bark conserves a tree's moisture content and insulates it so the moisture does not freeze. On the coldest days the wood beneath bark will feel damp.

Next year's leaves are already perfectly formed inside the small bud scales of the broad-leaved trees.

Late in winter, just before spring's arrival, tiny white blossoms called snowdrops pop up through the snow. They are found more often in clearings than in dense woodland.

Evergreens do not need to shed their needles in winter because less water escapes from them than from leaves of deciduous trees.

Beneath the frozen surface of ponds and streams the temperature is a constant 34.2°F. Even if a fish is trapped in the ice, it can live if a thaw comes before its body freezes.

Life goes on even in the snow. The tiny snow lice appear on warm days when the snow is thawing and feed on microscopic life on its surface.

The cardinal, once considered a southern bird, now winters farther north. Others that remain through the winter are chickadees, nuthatches, and sparrows.